Image/Video Coding and Processing Techniques for Intelligent Sensor Nodes

Image/Video Coding and Processing Techniques for Intelligent Sensor Nodes

Guest Editors

Jinjia Zhou
Ittetsu Taniguchi
Xin Jin

Basel • Beijing • Wuhan • Barcelona • Belgrade • Novi Sad • Cluj • Manchester

Guest Editors

Jinjia Zhou
Hosei University
Tokyo
Japan

Ittetsu Taniguchi
Osaka University
Osaka
Japan

Xin Jin
Tsinghua University
Shenzhen
China

Editorial Office
MDPI AG
Grosspeteranlage 5
4052 Basel, Switzerland

This is a reprint of the Special Issue, published open access by the journal *Sensors* (ISSN 1424-8220), freely accessible at: https://www.mdpi.com/journal/sensors/special_issues/Q78Y0U5HX0.

For citation purposes, cite each article independently as indicated on the article page online and as indicated below:

Lastname, A.A.; Lastname, B.B. Article Title. *Journal Name* **Year**, *Volume Number*, Page Range.

ISBN 978-3-7258-2865-4 (Hbk)
ISBN 978-3-7258-2866-1 (PDF)
https://doi.org/10.3390/books978-3-7258-2866-1

© 2024 by the authors. Articles in this book are Open Access and distributed under the Creative Commons Attribution (CC BY) license. The book as a whole is distributed by MDPI under the terms and conditions of the Creative Commons Attribution-NonCommercial-NoDerivs (CC BY-NC-ND) license (https://creativecommons.org/licenses/by-nc-nd/4.0/).

Contents

Jinjia Zhou, Ittetsu Taniguchi and Xin Jin
Image/Video Coding and Processing Techniques for Intelligent Sensor Nodes
Reprinted from: *Sensors* **2024**, *24*, 4819, https://doi.org/10.3390/s24154819 1

Dashan Zhang, Andong Zhu, Wenhui Hou, Lu Liu and Yuwei Wang
Vision-Based Structural Modal Identification Using Hybrid Motion Magnification
Reprinted from: *Sensors* **2022**, *22*, 9287, https://doi.org/10.3390/s22239287 3

Ionut Schiopu and Radu Ciprian Bilcu
Low-Complexity Lossless Coding of Asynchronous Event Sequences for Low-Power Chip Integration
Reprinted from: *Sensors* **2022**, *22*, 10014, https://doi.org/10.3390/s222410014 22

Sovann Chen, Supavadee Aramvith and Yoshikazu Miyanaga
Learning-Based Rate Control for High Efficiency Video Coding
Reprinted from: *Sensors* **2023**, *23*, 3607, https://doi.org/10.3390/s23073607 42

Chen Fu, Heming Sun, Zhiqiang Zhang and Jinjia Zhou
A Highly Pipelined and Highly Parallel VLSI Architecture of CABAC Encoder for UHDTV Applications
Reprinted from: *Sensors* **2023**, *23*, 4293, https://doi.org/10.3390/s23094293 62

Wenhao Wang, Zhenbing Liu, Haoxiang Lu, Rushi Lan and Yingxin Huang
Adapting Single-Image Super-Resolution Models to Video Super-Resolution: A Plug-and-Play Approach
Reprinted from: *Sensors* **2023**, *23*, 5030, https://doi.org/10.3390/s23115030 81

Jun Gong, Senlin Luo, Wenxin Yu and Liang Nie
Inpainting with Separable Mask Update Convolution Network
Reprinted from: *Sensors* **2023**, *23*, 6689, https://doi.org/10.3390/s23156689 99

Zheng Wang, Guancheng Quan and Gang He
Edge-Oriented Compressed Video Super-Resolution
Reprinted from: *Sensors* **2023**, *24*, 170, https://doi.org/10.3390/s24010170 119

Qinyu Wang, Chenxu Lu, Long Gao and Gang He
Transformer-Based Multiple-Object Tracking via Anchor-Based-Query and Template Matching
Reprinted from: *Sensors* **2023**, *24*, 229, https://doi.org/10.3390/s24010229 134

Zheng Wang and Gang He
Multi-Frame Content-Aware Mapping Network for Standard-Dynamic-Range to High-Dynamic-Range Television Artifact Removal
Reprinted from: *Sensors* **2024**, *24*, 299, https://doi.org/10.3390/s24010299 150

Kiho Choi
Block Partitioning Information-Based CNN Post-Filtering for EVC Baseline Profile
Reprinted from: *Sensors* **2024**, *24*, 1336, https://doi.org/10.3390/s24041336 166

Editorial

Image/Video Coding and Processing Techniques for Intelligent Sensor Nodes

Jinjia Zhou [1,*], Ittetsu Taniguchi [2] and Xin Jin [3]

1. Graduate School of Science and Engineering, Hosei University, Tokyo 102-8160, Japan
2. Graduate School of Information Science and Technology, Osaka University, Osaka 565-0871, Japan; i-tanigu@ist.osaka-u.ac.jp
3. Shenzhen International Graduate School, Tsinghua University, Shenzhen 518071, China; jin.xin@sz.tsinghua.edu.cn
* Correspondence: zhou@hosei.ac.jp

There is an increasing interest in developing intelligent sensor nodes which enable intelligent processing for Internet of Things (IoT) surveillance, remote sensing, and smart city applications. The data are processed on board by embedded signal processing and machine learning-based analysis algorithms. Such machine learning-driven sensors can transmit key information instead of raw sensing data, lowering the data volume traveling throughout the network.

Due to the explosion of image and video data in IoT systems, specifically designed image and video codecs have been preferred in recent years. With a focus on reducing the data burden and improving the reconstructed image quality, image/video coding and processing techniques performing in low-cost implementations, saving power consumption, and increasing the battery lifetime will cope with the design requirements of sensor nodes. Moreover, intelligent sensors change the traditional intuition-driven sensors in support of machine learning algorithms, delivering high-resolution images and videos for the 5G revolution.

In line with the mission of *Sensors*, the organizers of this Special Issue endeavor to apply the most recent advancements in image/video coding and processing techniques to the development of intelligent sensor nodes. (1) High compression ratio and high reconstruction quality algorithms are presented in this Special Issue. Kiho Choi proposed block partitioning information-based CNN post-filtering for an EVC baseline profile (Contribution 1). Sovann Chen et al. applied a learning-based rate control approach for High Efficiency Video Coding (HEVC) (Contribution 8). (2) Several low-cost video coding hardware designs are introduced. Chen Fu et al. implemented a highly pipelined and highly parallel VLSI architecture of a CABAC encoder for UHDTV applications (Contribution 7). This design can be used for multiple video coding standards such as H.264/AVC, H.265/HEVC, and H.266/VVC. Ionut Schiopu et al. developed a low-complexity lossless coding of asynchronous event sequences for low-power chip integration (Contribution 9). (3) To improve the quality of the compressed video, several techniques are presented in this Special Issue. Zheng Wang et al. introduced standard-dynamic-range to high-dynamic-range television artifact removal techniques by applying a multi-frame Content-Aware Mapping Network (Contribution 2). Zheng Wang proposed an edge-oriented compressed video super resolution that provides users with high-quality and cost-effective HR videos by integrating sensors and codecs (Contribution 4). Wenhao Wan et al. designed a super-resolution video algorithm by adapting single-image super-resolution models (Contribution 6). (4) Several vision-based tasks are presented. Qinyu Wang et al. introduced a transformer-based multiple-object tracking algorithm by applying an anchor-based-query and template matching (Contribution 3). Jun Gong proposed an inpainting algorithm with

Citation: Zhou, J.; Taniguchi, I.; Jin, X. Image/Video Coding and Processing Techniques for Intelligent Sensor Nodes. *Sensors* **2024**, *24*, 4819. https://doi.org/10.3390/s24154819

Received: 17 July 2024
Accepted: 22 July 2024
Published: 25 July 2024

Copyright: © 2024 by the authors. Licensee MDPI, Basel, Switzerland. This article is an open access article distributed under the terms and conditions of the Creative Commons Attribution (CC BY) license (https://creativecommons.org/licenses/by/4.0/).

a separable mask update convolution network (Contribution 5). Dashan Zhang et al. developed a vision-based structural modal identification using hybrid motion magnification (Contribution 10).

In summary, the papers featured in this collection present a variety of innovative approaches and solutions to the challenges in image/video coding and processing techniques for intelligent sensor nodes. The articles introduce algorithms, hardware designs, and deep learning-based strategies to address the development issues of intelligent image sensors. We extend our gratitude to the authors for their contributions and hope that this collection will inspire further advancements in the field of intelligent image/video sensors.

Conflicts of Interest: The authors declare no conflict of interest.

List of Contributions:

1. Choi, K. Block Partitioning Information-Based CNN Post-Filtering for EVC Baseline Profile. *Sensors* **2024**, *24*, 1336. https://doi.org/10.3390/s24041336.
2. Wang, Z.; He, G. Multi-Frame Content-Aware Mapping Network for Standard-Dynamic-Range to High-Dynamic-Range Television Artifact Removal. *Sensors* **2024**, *24*, 299. https://doi.org/10.3390/s24010299.
3. Wang, Q.; Lu, C.; Gao, L.; He, G. Transformer-Based Multiple-Object Tracking via Anchor-Based-Query and Template Matching. *Sensors* **2024**, *24*, 229. https://doi.org/10.3390/s24010229.
4. Wang, Z.; Quan, G.; He, G. Edge-Oriented Compressed Video Super-Resolution. *Sensors* **2024**, *24*, 170. https://doi.org/10.3390/s24010170.
5. Gong, J.; Luo, S.; Yu, W.; Nie, L. Inpainting with Separable Mask Update Convolution Network. *Sensors* **2023**, *23*, 6689. https://doi.org/10.3390/s23156689.
6. Wang, W.; Liu, Z.; Lu, H.; Lan, R.; Huang, Y. Adapting Single-Image Super-Resolution Models to Video Super-Resolution: A Plug-and-Play Approach. *Sensors* **2023**, *23*, 5030. https://doi.org/10.3390/s23115030.
7. Fu, C.; Sun, H.; Zhang, Z.; Zhou, J. A Highly Pipelined and Highly Parallel VLSI Architecture of CABAC Encoder for UHDTV Applications. *Sensors* **2023**, *23*, 4293. https://doi.org/10.3390/s23094293.
8. Chen, S.; Aramvith, S.; Miyanaga, Y. Learning-Based Rate Control for High Efficiency Video Coding. *Sensors* **2023**, *23*, 3607. https://doi.org/10.3390/s23073607.
9. Schiopu, I.; Bilcu, R.C. Low-Complexity Lossless Coding of Asynchronous Event Sequences for Low-Power Chip Integration. *Sensors* **2022**, *22*, 10014. https://doi.org/10.3390/s222410014.
10. Zhang, D.; Zhu, A.; Hou, W.; Liu, L.; Wang, Y. Vision-Based Structural Modal Identification Using Hybrid Motion Magnification. *Sensors* **2022**, *22*, 9287. https://doi.org/10.3390/s22239287.

Disclaimer/Publisher's Note: The statements, opinions and data contained in all publications are solely those of the individual author(s) and contributor(s) and not of MDPI and/or the editor(s). MDPI and/or the editor(s) disclaim responsibility for any injury to people or property resulting from any ideas, methods, instructions or products referred to in the content.

Article

Vision-Based Structural Modal Identification Using Hybrid Motion Magnification

Dashan Zhang [1,2], Andong Zhu [1,2], Wenhui Hou [1,2], Lu Liu [1,2] and Yuwei Wang [1,2,*]

1. College of Engineering, Anhui Agricultural University, Hefei 230036, China
2. Anhui Province Engineering Laboratory of Intelligent Agricultural Machinery and Equipment, Anhui Agricultural University, Hefei 230036, China
* Correspondence: wyw@ahau.edu.cn; Tel.: +86-150-5607-2806

Abstract: As a promising alternative to conventional contact sensors, vision-based technologies for a structural dynamic response measurement and health monitoring have attracted much attention from the research community. Among these technologies, Eulerian video magnification has a unique capability of analyzing modal responses and visualizing modal shapes. To reduce the noise interference and improve the quality and stability of the modal shape visualization, this study proposes a hybrid motion magnification framework that combines linear and phase-based motion processing. Based on the assumption that temporal variations can represent spatial motions, the linear motion processing extracts and manipulates the temporal intensity variations related to modal responses through matrix decomposition and underdetermined blind source separation (BSS) techniques. Meanwhile, the theory of Fourier transform profilometry (FTP) is utilized to reduce spatial high-frequency noise. As all spatial motions in a video are linearly controllable, the subsequent phase-based motion processing highlights the motions and visualizes the modal shapes with a higher quality. The proposed method is validated by two laboratory experiments and a field test on a large-scale truss bridge. The quantitative evaluation results with high-speed cameras demonstrate that the hybrid method performs better than the single-step phase-based motion magnification method in visualizing sound-induced subtle motions. In the field test, the vibration characteristics of the truss bridge when a train is driving across the bridge are studied with a commercial camera over 400 m away from the bridge. Moreover, four full-field modal shapes of the bridge are successfully observed.

Keywords: operational modal analysis; vision-based measurement; hybrid motion magnification; modal shapes visualization; temporal and spatial denoising

1. Introduction

Structural experimental modal parameters, including modal frequencies, damping ratios, and modal shapes, provide insight into dynamic behaviors and are critical for applications such as structural health monitoring (SHM) and nondestructive testing (NDT) [1]. Usually, these properties are recovered by analyzing the vibrations of limited discrete points on the object through reliable contact sensors. However, the physically attached sensors may cause a mass-loading effect on lightweight targets, and they are difficult to affix to complex large-scale structures [2–4]. As an alternative, the vision-based method is one of the most popular non-contact measurement methods for the structural modal analysis in recent years [5–7]. Compared with common contact sensors, camera-based devices are more flexible and provide a higher spatial-resolution sensing capacity, which makes them convenient for remote installation and preferable for full-field measurements [8–10].

With advances in image processing techniques (e.g., image registration and optical flow), vision-based measurements can obtain intuitionistic image sequences of structural vibrations and are applied to experimental modal tests for various types of structures. In most cases [11,12], these methods extract the field deformation or local displacement from variations in speckle and high-contrast natural or artificial makers on the surface of the structure, which limits their applications to featureless and large-scale measurements. Meanwhile, although subpixel precision can be achieved, for extremely subtle motions, it is still difficult for algorithms to balance the efficiency and resolution, especially when full-field measurements are required.

As a computation technique for visualizing subtle colors and variations in videos, Eulerian video magnification [13–15] shows a strong vitality and is used in actual output-only modal analyses [16–18]. Unlike the motion extraction approaches based on an interframe correlation or gradient, the Eulerian approach considers that the structural in-plain small motion is closely related to the intensity or phase variations in the timeline. A quantitative analysis of these temporal variations reveals vital characteristics (e.g., the elasticity and modal frequency). Moreover, by manipulating the spatial motion, structural modal shapes can be directly observed in motion-magnified videos [19–24].

Although magnifying the temporal intensity or phase variations can achieve motion magnification, these two frameworks have different characteristics. Normally, linear processing is more sensitive to subtle motions and less robust to the noise from imaging sensors and illumination. Meanwhile, phase-based motion processing performs better in noise control and can support a larger amplification factor, so it is more suitable for visualizing and understanding modal shapes. However, both spatial and temporal noises severely affect the quality of the outputs of phase-based motion processing, especially in a subtle and long-distance motion observation [18–20,24,25]. In practice, it is difficult to uniformly reduce the temporal phase noise without any prior information of the measured structures. In addition, the existence of multi-scale decomposition also increases the complexity of the noise processing in both space and timeline [14,18,23].

To reduce noises and improve the quality of the modal shape visualization, it is desirable to propose a hybrid motion magnification framework that combines linear and phase-based motion processing. Based on the assumption that temporal variations can approximate spatial motions, previous studies [26] have shown that the singular value decomposition (SVD) can extract the structural vibration from the temporal intensity variations, and the spatial motions in videos can be manipulated linearly with a higher efficiency. Considering that the extracted temporal variations relevant to vibrations are mixed signals, the sparse component analysis (SCA) technique is used in signal separation [27,28]. Meanwhile, as noises mainly exist in the high-spatial-frequency part [13–15], Fourier transform profilometry (FTP) is utilized to improve the weights that represent the severity of the spatial motion [29,30]. In the hybrid framework, linear motion processing simplifies the processes of vibration extraction and noise reduction and provides high-quality, controllable inputs for visualizing modal shapes in phase-based motion processing. The proposed framework was applied to two laboratory experiments and a field test on a large-scale truss bridge to evaluate its performance in a modal analysis.

The main contributions of this paper are summarized as follows: (1) A linear motion processing approach is proposed to extract and manipulate the structural vibrations in videos. Meanwhile, a set of methods is developed to reduce the temporal and spatial noises. (2) The high-quality visualization of structural modal shapes is realized in the hybrid motion magnification framework. (3) The performance of the proposed framework is investigated through sound-induced modal tests in the laboratory. The effectiveness of this proposed framework is verified in a long-distance field test to analyze the vibration characteristics of a large-scale truss bridge.

The rest of this paper is organized as follows. Section 2 introduces the proposed hybrid motion magnification framework, including the details of the temporal and spatial noise reduction. The experimental data with a lightweight beam from the MIT CSAIL [19] are analyzed to better understand the implementation scheme. Section 3 validates the proposed method through a set of experiments and discusses its advantages and limitations. Section 4 concludes this paper.

2. Materials and Methods

2.1. Structural Vibration and Intensity Variations

Modal analysis models a solid object as a system of point masses connected by springs and dampers. Without loss of generality, the differential equation of a multi-DOF vibration system is expressed as

$$\mathbf{M}\ddot{p} + \mathbf{C}\dot{p} + \mathbf{K}p = 0, \tag{1}$$

where $\mathbf{M}$ is the mass matrix; $\mathbf{C}$ and $\mathbf{K}$ are matrices describing the viscous damping values and spring stiffness between points, respectively; p, $\dot{p}$, and $\ddot{p}$ are vectors indicating the displacement, velocity, and acceleration of the points, respectively. Under the assumption of Rayleigh damping, matrix $\mathbf{C}$ is ideal and is assumed to be a linear combination of $\mathbf{M}$ and $\mathbf{K}$. After the generalized eigenvalue problem is solved, the system can be decoupled into single-degree freedom systems, and the vibration motion of modal masses can be expressed as a linear combination of the modal responses:

$$\mathbf{p}(t) = \Phi \mathbf{q}(t) = \sum_{i=1}^{n} \phi_i q_i(t), \tag{2}$$

where n is the mode number; Φ is the modal matrix that defines modal coordinates $\mathbf{q}(t)$; ϕ_i and q_i are, respectively, the i-th mode shape and modal coordinate.

With the assistance of imaging, structural vibration can be measured by video records containing frames with temporally translated image intensities. From the Eulerian perspective, temporal filtering can approximate spatial translation [13–15,19]. To investigate the relationship between intensity variations and vibration, the simple case of 1D signal translation is considered in this paper. Let $I(x,t)$ be the image intensity at position x and time t. The observed intensity variations can be expressed by a displacement function $\delta(x,t)$, and $\delta(x,0) = 0$. All valuable intensity variations $\tilde{\delta}(x,t)$ should be highly correlated with the modal responses:

$$\tilde{\delta}(x,t) := \sum_{i=1}^{n} w_i(x) q_i(t), \tag{3}$$

where $w_i(x)$ is the weight corresponding to the i-th mode shape (related to modal coordinates). Thus, the displacement function $\delta(x,t)$ is expressed as the combination of $\tilde{\delta}(x,t)$ and noise:

$$\delta(x,t) = \sum_{i=1}^{n} w_i(x) q_i(t) + N(x,t), \tag{4}$$

where $N(x,t)$ is the noise mainly caused by the environment and imaging.

2.2. Linear Motion Processing

From Equation (4), to achieve linear motion magnification at a particular resonant frequency, $\tilde{\delta}(x,t)$ needs to be estimated and decoupled, and the noise $N(x,t)$ needs to be reduced at every pixel coordinate:

$$\tilde{I}_i(x,t) = I(x,t) + \alpha_i(w_i(x) q_i(t) + n_i(x,t)), \tag{5}$$

where α_i is the amplification factor for the i-th mode, and $n_i(x,t)$ is the residual noise ($n_i(x,t) \ll N(x,t)$).

Based on the assumption that useful intensity variations and noises are linearly independent, $\tilde{\delta}(x,t)$ and $N(x,t)$ can be separated by using SVD efficiently [26]. For each pixel coordinate, the difference between $I(x,t)$ and $I(x,0)$ is calculated and then used to reshape matrix $\mathbf{D}$ that represents the temporal intensity variations in the video. Through SVD, this matrix is decomposed, and k significant singular values are reserved:

$$\mathbf{D}_{c \times l} \xrightarrow{SVD} \mathbf{U}_{c \times k} \cdot \mathbf{S}_{k \times k} \cdot \mathbf{V}^{*}_{l \times k} = \sum_{r=1}^{k} \mathbf{u}_r s_r \mathbf{v}_r^{*}, \tag{6}$$

where c is the total number of pixel coordinates; l is the length of the video; s_r is reserved singular value; $\mathbf{u}_r$ and $\mathbf{v}_r$ are, respectively, orthogonal left-singular and right-singular vectors; symbol $*$ means matrix transposition. The reserved $s_r \mathbf{v}_r^*(r=1,2,...,k)$ are considered as the output observations representing an instantaneous linear mixture of signals $q_i(t)(i=1,2,...,n)$:

$$\begin{bmatrix} s_1 v_1^* \\ s_2 v_2^* \\ \vdots \\ s_k v_k^* \end{bmatrix} = \mathbf{A} \begin{bmatrix} q_1(t) \\ q_2(t) \\ \vdots \\ q_n(t) \end{bmatrix}, \quad (7)$$

where $\mathbf{A}$ is referred to as the mixing matrix. The reserved four temporal intensity variations, their frequency spectra, and the corresponding weights are illustrated in Figure 1. It can be seen that the reserved intensity variations are coupled signals of multiple modal responses [19].

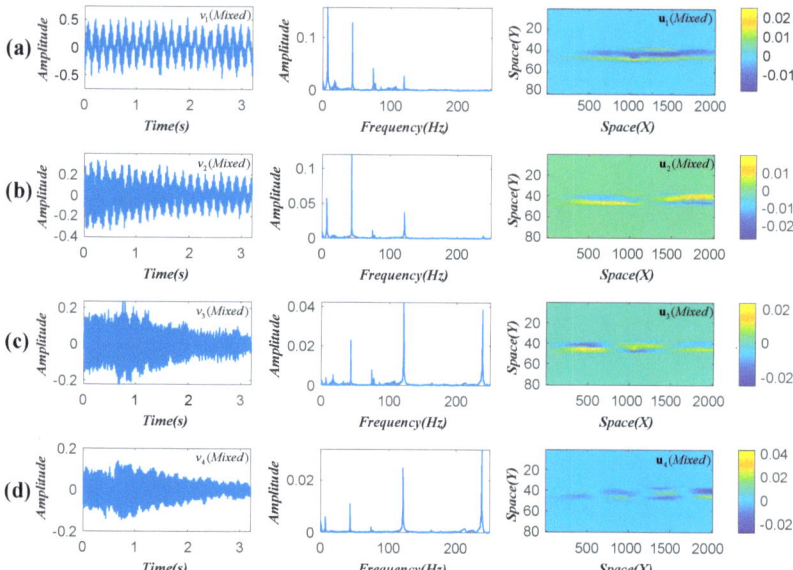

Figure 1. Reserved temporal intensity variations and corresponding weights in beam test (**a–d**).

Taking Equation (7) as an operational modal analysis (OMA) problem, the mixing matrix can be estimated by using the blind source separation (BSS) technique. The well-posedness of Equation (7) is determined by the magnitude of k (the number of reserved singular values) and n (the activated maximum mode order). In this paper, the equation is considered an underdetermined BSS problem, and mixing matrix $\mathbf{A}$ is estimated via SCA [27,28]. By decoupling the reserved intensity variations, the corresponding weights are updated as follows:

$$\tilde{\mathbf{u}}_i = \sum_{r=1}^{k} \mathbf{u}_r s_r \mathbf{v}_r^* \times q_i(t)^*. \quad (8)$$

Thus, according to Equation (5), the output of linear motion processing can be expressed as

$$\tilde{I}_i(x,t) = I(x,t) + \alpha_i(\tilde{\mathbf{u}}_i(x)q_i(t)) \approx I(x,0) + \sum_{i=1}^{n} w_i(x)q_i(t) + N(x,t) + \alpha_i(\tilde{\mathbf{u}}_i(x)q_i(t)). \quad (9)$$

To further reduce the noises and remove the existing vibrations in the input video, this process can be performed on the first frame of the video sequence, i.e.,

$$\tilde{I}_i(x,t) = I(x,0) + \alpha_i(\tilde{\mathbf{u}}_i(x)q_i(t)). \tag{10}$$

Figure 2a shows the scatter diagram of the first three temporal intensity variations $(s_r \mathbf{v}_r^*(r = 1, 2, 3))$. The modal assurance criterion (MAC) in Figure 2b is used to determine the errors of the estimated mode shape vectors. The observed directions in Figure 2a represent the estimated four mode shape vectors, and the theoretical mode shape vectors are calculated by using the FEM software. The time-domain modal responses are recovered by using the l_1-optimization algorithm [27]. The decoupled temporal intensity variations, their frequency spectra, and the updated weights are illustrated in Figure 3. It is considered that these decoupled temporal intensity variations are highly correlated with the first four modal responses [19].

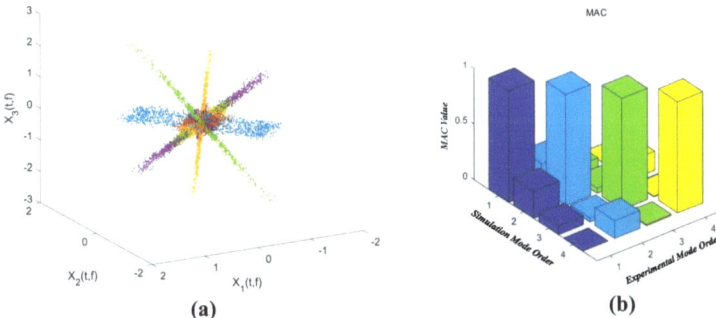

Figure 2. (**a**) Scatter diagram of the first three measuring signals and (**b**) the modal assurance criterion (MAC).

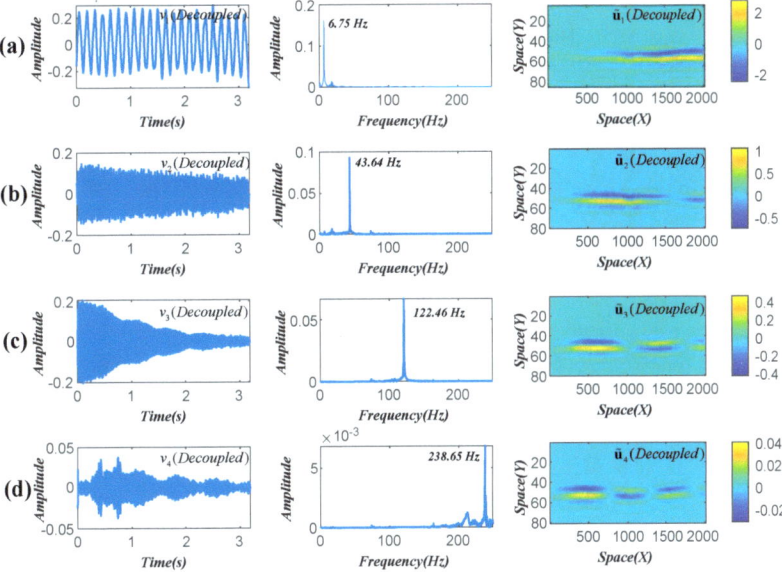

Figure 3. Decoupled temporal intensity variations and corresponding weights in beam test. (**a**) First mode; (**b**) second mode; (**c**) third mode; and (**d**) fourth mode.

2.3. Weight Enhancement of the FTP

According to Equation (8), the spatial weights $\tilde{u}_i$ are calculated by using the decoupled intensity variations. As the linear motion processing above does not consider the spatial consistency, the updated weights $\tilde{u}_i$ are not spatially smooth and continuous. As noises mainly exist in high spatial-frequency temporal variations, the FTP is utilized to improve the quality of spatial weights [29,30].

Taking $\tilde{u}_1$ as an example, Figure 4 shows the enhancement process. Let the spatial weights $\tilde{u}_1$ (Figure 4a) deform the reference grating image, the deformed grating image is shown in Figure 4b, and the spatial-frequency spectra of the deformed grating image are shown in Figure 4c. Assuming that the noise is mainly related to the high-frequency component in the spatial-frequency spectra, the spectra are filtered to let only the fundamental component through (red circle), and then reversed Fourier transform is applied to the fundamental component. According to the theory of FTP, the core variable that varies directly with the spatial weights is the phase distribution. The formula to obtain the improved spatial weights is given as

$$\hat{u}_i = \frac{l_0 \Delta \eta_i}{\Delta \eta_i - 2\pi f_0 d}, \quad (11)$$

where $\Delta \eta_i$ is the unwrapped phase difference; f_0 is the fundamental frequency of the observed grating image; l_0 and d are preset values in the crossed-optical-axes geometry of FTP. The improved spatial weights $\hat{u}_1$ are shown in Figure 4d.

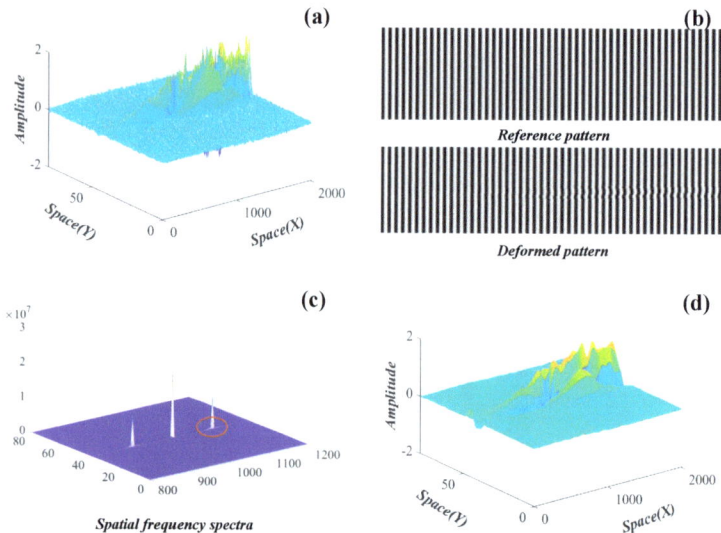

Figure 4. The spatial weight enhancement process of $\tilde{u}_1$. (**a**) The spatial weight $\tilde{u}_1$; (**b**) the reference and deformed grating images; (**c**) the spatial-frequency spectra of the deformed grating image; (**d**) the improved spatial weight $\hat{u}_1$.

The results and analyses of the improved spatial weights in the beam test are presented in Figures 5 and 6. Subfigures (1) and (2) in Figures 5 and 6 compare the original and the improved spatial weights. It can be seen that the spatial weights improved by FTP are much smoother than the originals and have better performance on spatial consistency. Subfigures (3) and (4) in Figures 5 and 6 compare the sampling results of the original and the improved weights in different spatial directions (the red and yellow lines in subfigure (1)). The results indicate that the noises in improved spatial weights are significantly reduced.

According to Equation (10), linear motion magnification can be achieved by using the decoupled temporal intensity variations and improved spatial weights. Figure 7 illustrates the linear motion magnification results in the beam test of MIT CSAIL [19]. The effectiveness of linear motion processing is validated with the spatiotemporal pixel slices cut from the motion-magnified videos. The mean intensity values inside the green circle in the background are calculated to study the residual noise. The analysis results indicate that these motion-magnified videos obtained by using the improved spatial weights achieve better performance on noise control.

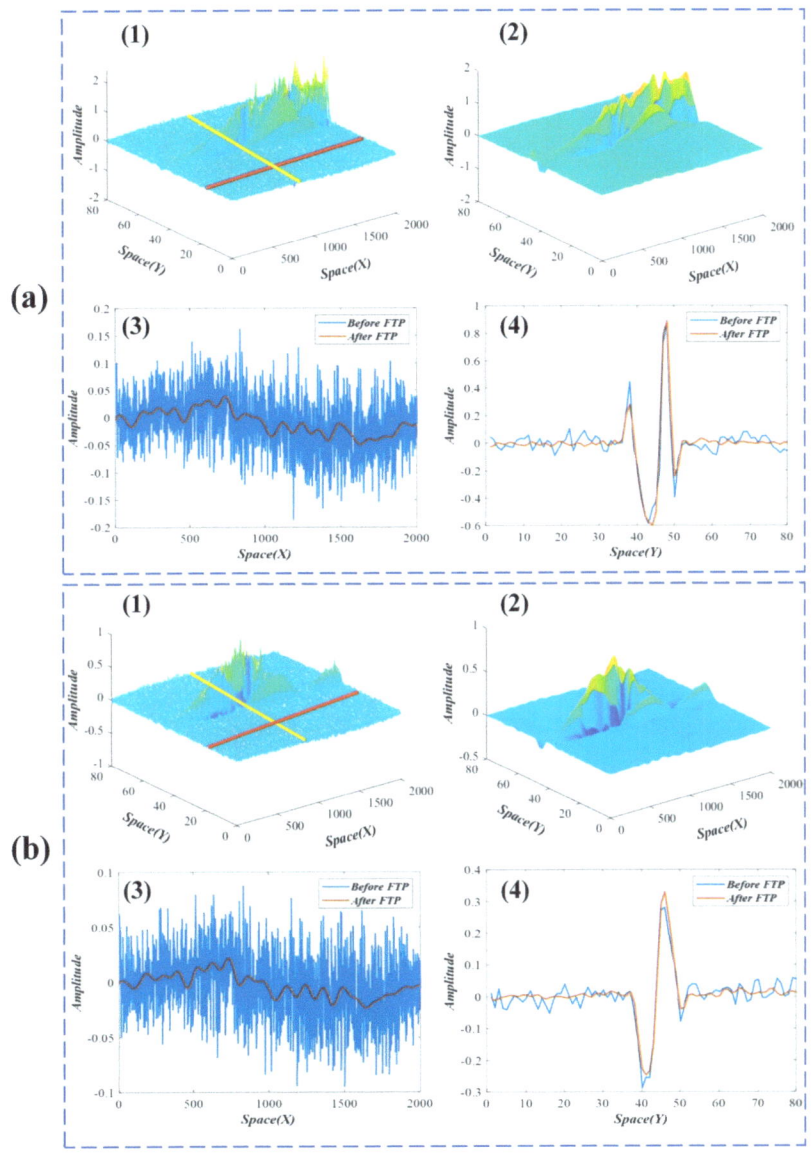

Figure 5. (**a**,**b**), the FTP results and analyses of the first two modes in the beam test. (1) Original spatial weights; (2) improved spatial weights; (3) and (4), sampling results of the original and the improved weights in x and y directions.

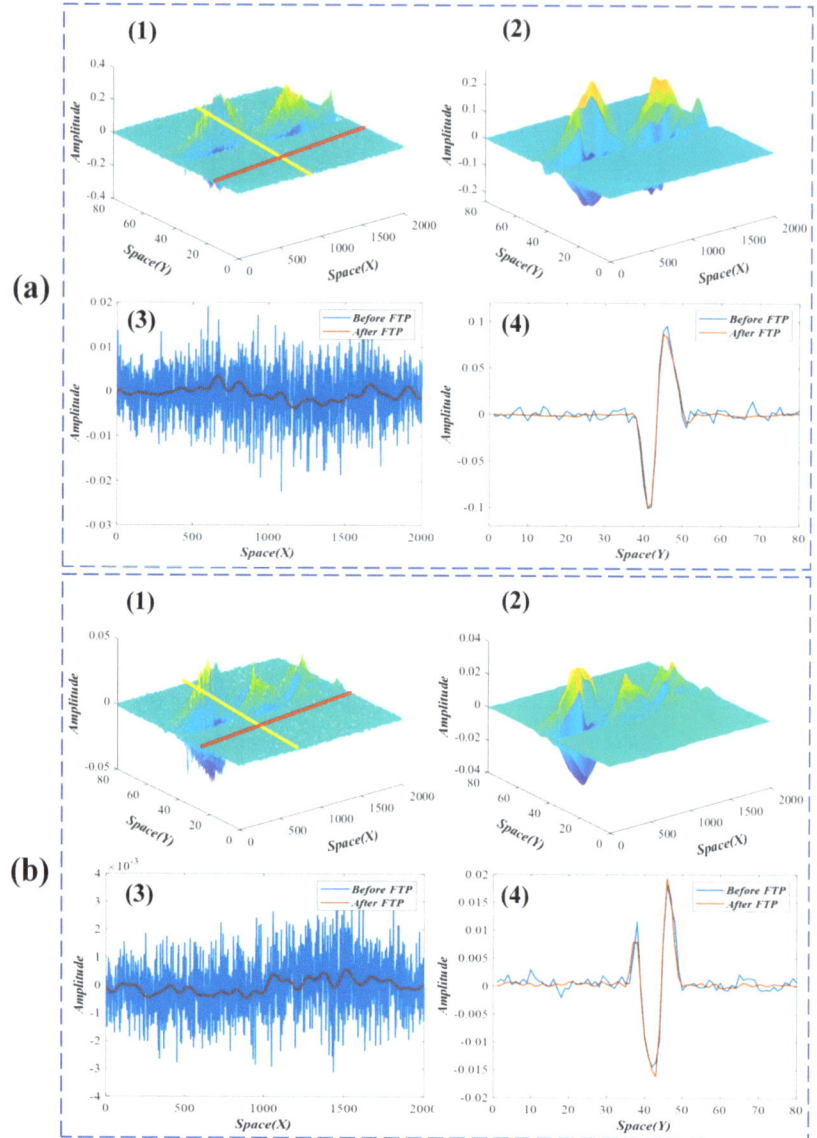

Figure 6. (**a**,**b**), the FTP results and analyses of the third and fourth modes in the beam test. (1) Original spatial weights; (2) improved spatial weights; (3) and (4), sampling results of the original and the improved weights in x and y directions.

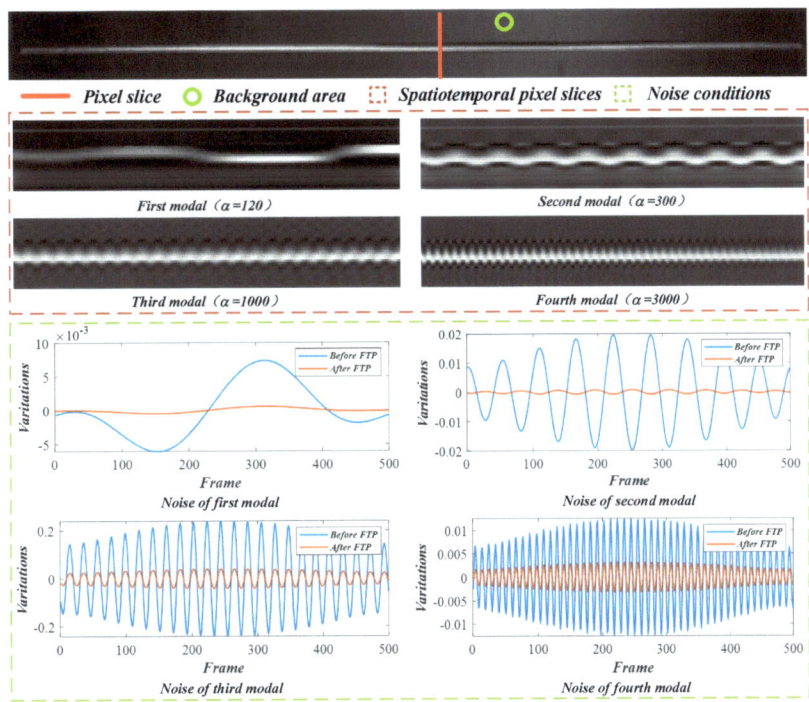

Figure 7. Linear motion processing results and noise reduction analysis in the beam test.

2.4. Phase-Based Motion Processing

The reason for further phase-based motion processing is that this framework can support large amplification factors and show better noise performance than the linear approximation. From Equation (10), the structural vibrations in the video can be initially produced through linear motion processing. Based on phase-based motion processing, the produced image profile can be decomposed into a sum of complex sinusoids by using the Fourier series:

$$\tilde{I}_i(x,t) = \sum_{\omega=-\infty}^{\infty} B_\omega e^{j\omega(x+\alpha_i \hat{u}_i(x) q_i(t))}. \tag{12}$$

Let $\tilde{\delta}_i(x,t) = \hat{u}_i(x) q_i(t)$ denote the initial motion. The band corresponding to a single frequency ω is the complex sinusoid:

$$S_\omega(x,t) = B_\omega e^{j\omega(x+\alpha_i \tilde{\delta}_i(x,t))}. \tag{13}$$

Because the initial motions in the video are controlled according to specific mode shapes, in phase-based motion processing, only the temporal DC component [14] of the phase $\omega(x+\alpha_i \tilde{\delta}_i(x,t))$ needs to be removed. Then, the phase $\alpha_i \tilde{\delta}_i(x,t)$ is multiplied with another amplification factor β_i to obtain the motion magnified sub-band:

$$\tilde{S}_\omega(x,t) = S_\omega(x,t) e^{j\alpha_i \beta_i \omega \tilde{\delta}_i(x,t)} = B_\omega e^{j\omega(x+(1+\beta_i)\alpha_i \tilde{\delta}_i(x,t))}. \tag{14}$$

The motion-magnified sequence can be eventually reconstructed by summing all the sub-bands. The total magnification factor in Equation (14) is $(1+\beta_i)\alpha_i$.

The motion magnification results of the original phase-based method and our improved framework are compared, and the result is shown in Figure 8 (8 orientations, half-octave bandwidth pyramids). The filter bands of the original phase-based approach are set to ±2 Hz near the experimental modal frequencies of the test beam. Table 1 presents the magnification factors and compares the image quality results of the reconstructed videos. The results of the average blind/referenceless image spatial quality evaluator (BRISQUE) [31] indicate that these videos reconstructed by the improved framework have better image quality. The average BRISQUE score of the input video image is 41.98. Because the initial motion is 0 (Equation (10)), the overall amplification factors of our improved framework are larger than those used in the original.

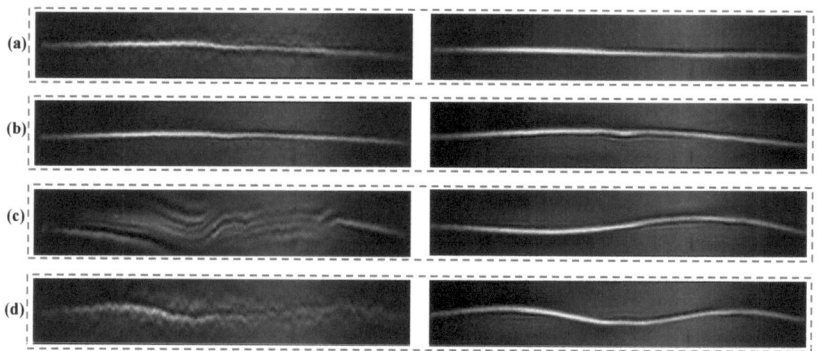

Figure 8. The comparison between the original phase-based method and our improved method. (**a**) The first mode; (**b**) the second mode; (**c**) the third mode; and (**d**) the fourth mode. More results are shown in the Supplementary videos.

Table 1. The amplification factors and image quality in the beam test.

Mode order	Factor (Original)	Factor (α_i)	Factor (β_i)	BRISQUE (Original)	BRISQUE (Improved)
1st	350	10	40	50.94	45.48
2nd	600	30	25	46.78	44.13
3rd	1000	100	15	50.20	44.30
4th	12,000	1000	20	56.00	43.69

3. Results

3.1. Vibration Analysis of a Lightweight Beam

In the first case, the modal parameters of a lightweight beam from a video are analyzed in a controlled laboratory experiment. The experiment setup is illustrated in Figure 9. The lightweight beam made of alloy steel was clamped with a table vice. During the experiment, audio with a frequency band ranging from 10 to 500 Hz was played by the loudspeaker about 0.1 m away from the surface of the beam at 80 decibels. When air fluctuations reach the beam, subtle forced vibrations will appear on the surface. Meanwhile, subtle vibrations motivated by the excitation audio were recorded by the high-speed camera system (Revealer 5KF10M, Agile Device Inc., Hefei, China) at 500 fps with a resolution of 580 × 180 pixels. The dimension and the material parameters of the beam are listed in Table 2. According to the Euler–Bernoulli beam theory, the theoretical resonant frequencies of a cantilever beam are estimated as follows:

$$f_n = \frac{3.52\gamma}{2\pi l^2}\sqrt{\frac{ER}{\rho A}}, (\gamma = 1, 6.27, 17.55, 34.39\ldots), \tag{15}$$

where f_n, E, and R denote the resonant frequency of the n-th mode, Young's Modulus, and the moment of inertia of the beam, respectively; ρ, A, and l denote the density, the cross-sectional area, and the length of the beam, respectively.

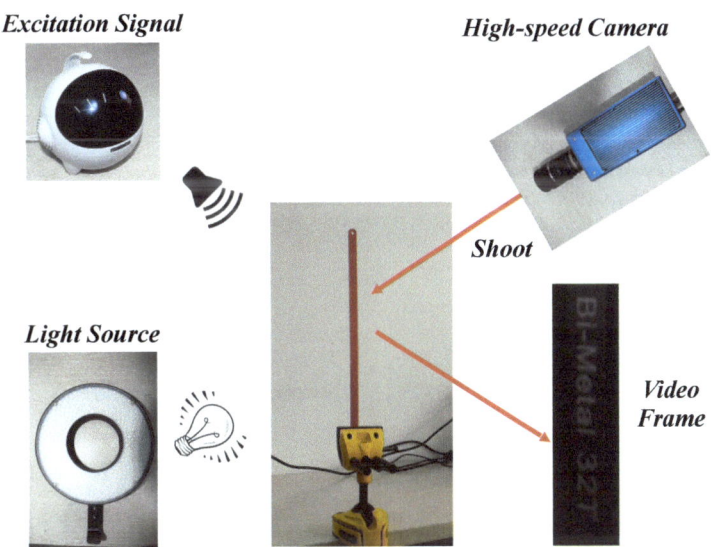

Figure 9. Experiment setup of the laboratory lightweight beam test.

Table 2. Parameters of the lightweight beam.

Dimensions (mm)	Young's Modulus	Density
$290 \times 12.6 \times 0.65$	2.06×10^{11} N·m^{-2}	7.85×10^{3} kg·m^{-3}

After the SVD decomposition, two temporal intensity variations were reserved. Their waveforms, frequency spectra, and the corresponding weights are shown in Figure 10. By decoupling the two reserved intensity variations through an SCA, four obvious peaks, including 6.37, 40.16, 113.10, and 221.60 Hz, were detected from the power spectra of the decoupled signals. According to Equation (15), these four temporal intensity variations are connected with the subtle spatial motions of the first four modal shapes. The comparison between the theoretical and experimental resonant frequencies is illustrated in Table 3. The decoupled intensity variations, their frequency spectra, and the updated weights (enhanced by FTP) are shown in Figure 11.

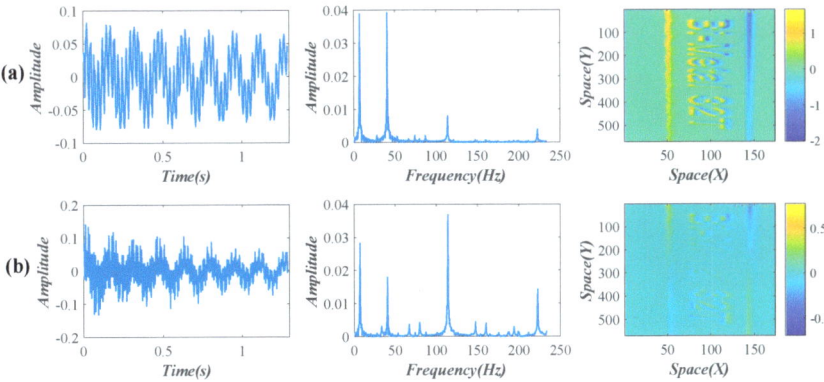

Figure 10. (**a**,**b**), reserved temporal intensity variations, their frequency spectra, and the corresponding weights in the lightweight beam test.

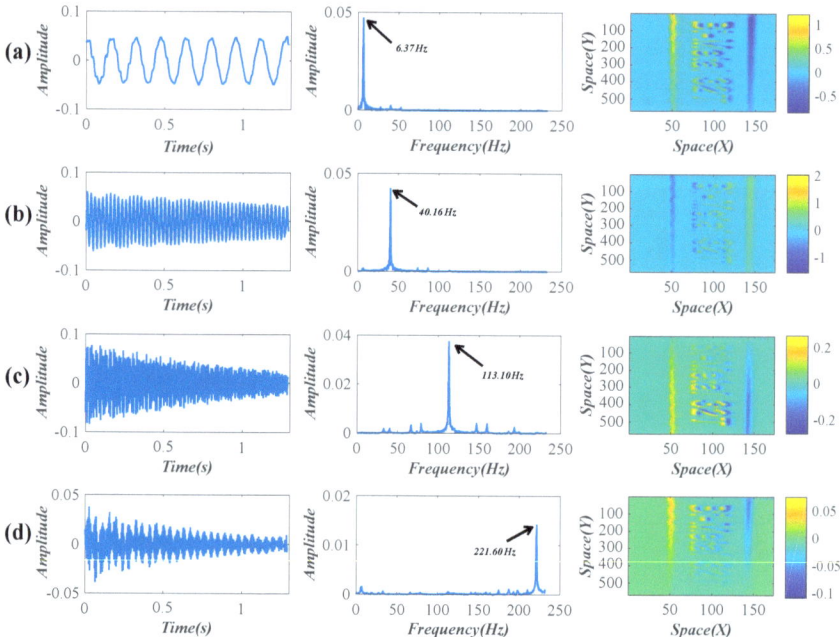

Figure 11. (a–d) Decoupled temporal intensity variations, their frequency spectra, and the corresponding weights (enhanced by FTP) in the lightweight beam test.

Table 3. Comparison between the theoretical and experimental resonant frequencies in the lightweight beam test.

Mode order	Theoretical (Hz)	Experimental (Hz)	Error Rate (%)
1st	6.41	6.37	0.62
2nd	40.17	40.16	0.02
3rd	112.43	113.10	0.59
4th	220.31	221.60	0.08

After the decoupled temporal intensity variations and the corresponding spatial weights were obtained, the motions in the video frames can be produced linearly and then magnified through the phase-based processing. The complex steerable pyramids (eight orientations, half-octave bandwidth pyramids) were used to decompose the video frames, and the local phases in different spatial scales and orientations over time were obtained. The filter bands for the original phase-based magnification were set to ± 2 Hz near the experimental modal frequencies. Figure 12 compares the final motion-magnified videos obtained by the original and our improved framework. The four colored lines in Figure 12a indicate the locations of the spatiotemporal pixel slices. Figure 12b–e) show the spatiotemporal slices of the first to the fourth modal shapes, respectively. It can be seen that the beam in the video reconstructed by our framework (solid line boxes) vibrates properly following a specific vibration mode, and the existing motions in the input video are removed. Table 4 presents the magnification factors and compares the image quality of the processed videos. The average BRISQUE score of the input video image is 39.34. The motion-magnified videos of the improved framework achieve a better image quality.

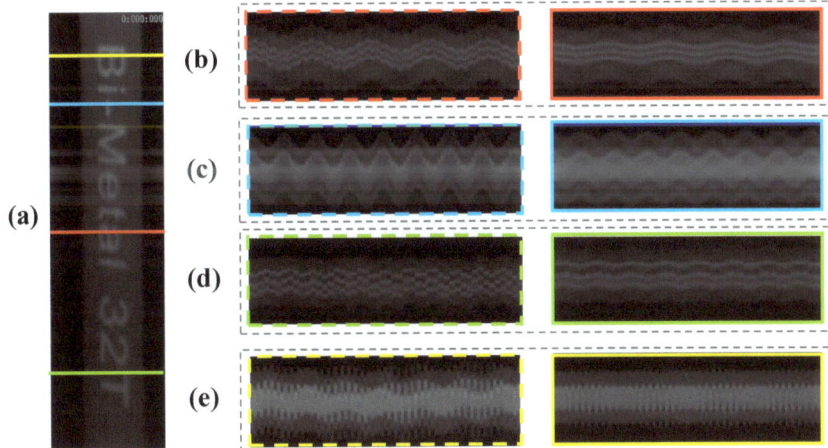

Figure 12. (a) Video frame and location of spatiotemporal slices and (**b–e**) spatiotemporal slices comparisons between the original and our improved framework in the lightweight beam test. More results are shown in the Supplementary videos.

Table 4. The amplification factors and image quality in the case of the lightweight beam.

Mode Order	Factor (Original)	Factor (α_i)	Factor (β_i)	BRISQUE (Original)	BRISQUE (Improved)
1st	10	0.15	100	50.84	40.45
2nd	15	0.2	200	52.10	41.58
3rd	60	2	50	48.04	41.75
4th	100	5	30	49.24	40.78

3.2. Vibration Analysis of the Nanfeihe Truss Bridge

The modal parameters of bridges reflect their vibration characteristics and are significant for bridge design and a structural state assessment. For large-scale bridges, it is difficult to excite the heavy structure with traditional vibration excitation devices and obtain structural vibration modes. In the second experiment, the vibration of the Nanfeihe railway truss bridge is observed under the wind–train–bridge coupling condition with a commercial camera, and the four modal shapes of the bridge are visualized with our improved motion magnification framework.

The Nanfeihe railway truss bridge is a super-spanned railway bridge about 8 km away from Hefei South Railway Station. The bridge is a low-supported steel bridge composed of a continuous truss and flexible arch with a main span of 229.5 m. The rise of the arch is 45 m, and the overall length of the bridge is 461 m. The weight of the bridge is over 13,000 tons. As shown in Figure 13, a commercial camera (Canon 70D) was installed about 410 m away from the center of the bridge by the Nanfei river. Figure 13a illustrates the satellite view of the camera measurement location relative to the bridge. The camera and the bridge in the same view are shown in Figure 13b. During the filming, a high-speed train was driving across the bridge. The camera recorded the whole process at 25 fps with the resolution of 1920 × 1080 pixels. A 645 × 1775 pixel region of interest (ROI) was selected from the screenshot to reduce the interference of the background (Figure 13c).

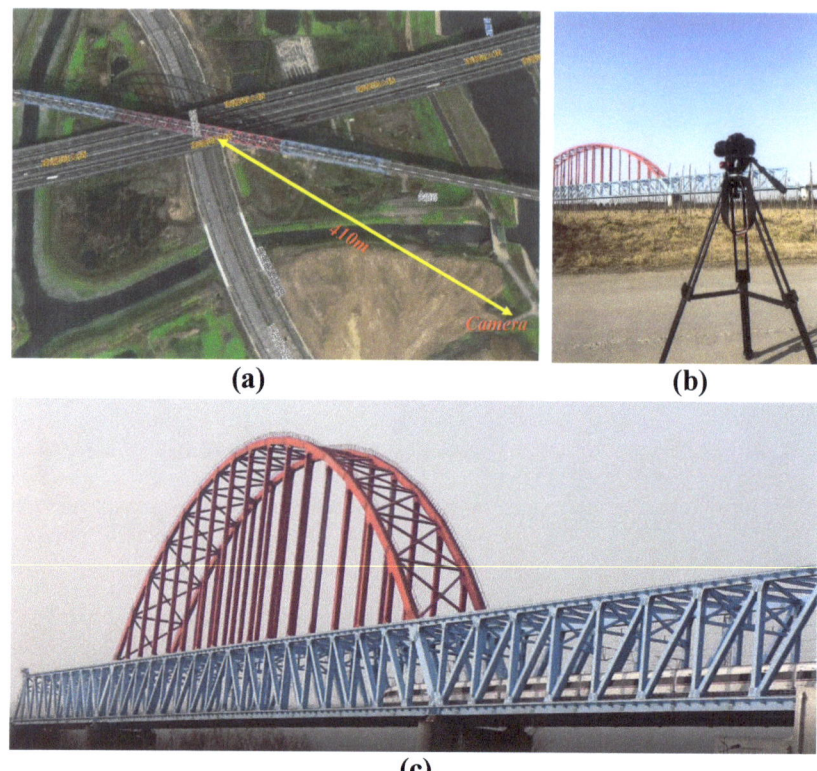

Figure 13. (**a**) The distance between camera and center of the bridge (image ©Baidu), (**b**) camera system, and (**c**) selected ROI.

For the data analysis, it is necessary to discuss and specify the vibration situation of the truss bridge. Before the high-speed train arrives, the vibration of the bridge is mainly caused by the environmental wind load. When the train arrives on the bridge, deflection appears on the structure, and the bridge will be affected by the excitation of both the wind and the train. For simplicity, this paper assumes that the two vibration processes (forced by the wind and the wind and train) are steady, compelled vibration processes and then discusses the influence of the transient vibration.

Figure 14 shows the two reserved temporal intensity variations after the SVD decomposition of the pixel difference matrix from the video file. As shown in the first column in Figure 15, four independent components (red curves) are decoupled from the reserved variations through an SCA. When the train arrives on the bridge, large variations that reflect the bridge deflections appear on the curves. The train arrival time and train leaving time are marked by arrows in Figure 15a and are found at corresponding positions in Figure 15b–d. To investigate the influence of deflections, this paper detrends these intensity variations (blue curves) and then separates the signals according to the difference in the excitation source before a frequency analysis. The second and the third columns in Figure 15 show the power spectra of the detrended intensity variations before and after the train arrives. The vibration frequency under the load of the wind is 0.78 Hz, and the main frequency under the excitation of the train is 4.19 and 8.39 Hz (a multiple frequency of 4.19 Hz) [32–34]. The updated spatial weights are illustrated in the last column of Figure 15. From the angle perpendicular to the image plane, these spatial weights exhibit the four different vibration modes of the bridge. It is worth mentioning the FTP was not utilized here because the frequency distribution of the spatial weights was too complex to be separated by the FTP.

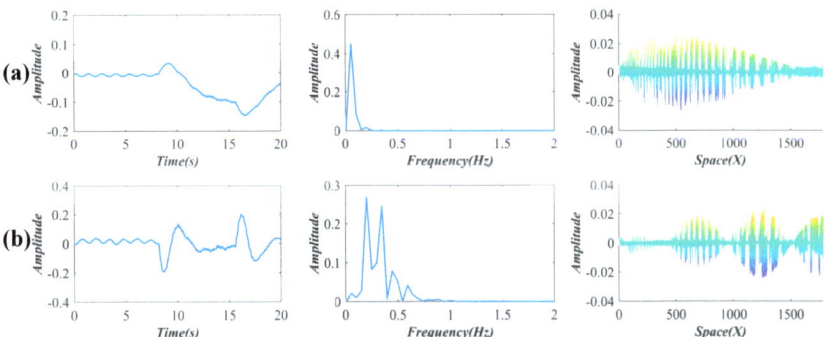

Figure 14. (**a**,**b**), reserved temporal intensity variations, their frequency spectra, and the corresponding weights in the bridge test.

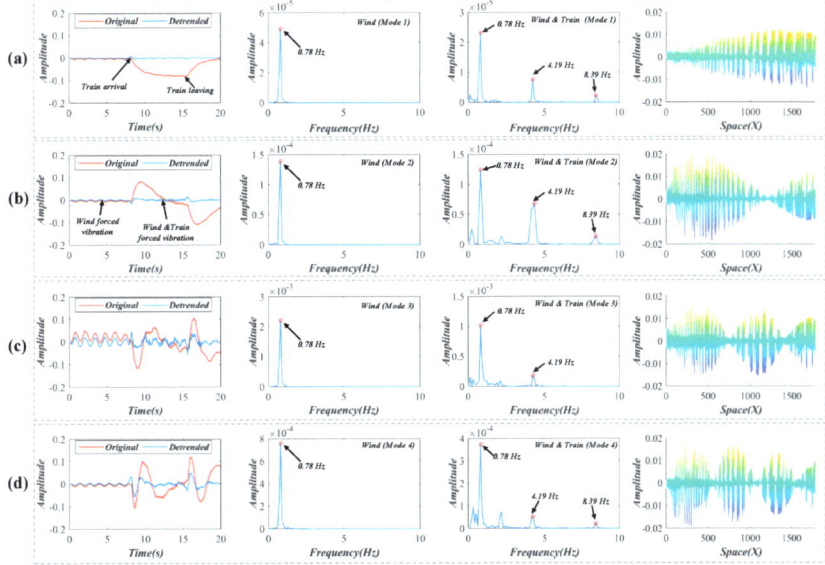

Figure 15. (**a**–**d**) Decoupled temporal intensity variations, power spectra under different excitations and weights in the bridge test. The first column represents the original and detrended reserved variations. The second and third columns show the power spectra before and train arrival time. The last column shows the decoupled weights in the bridge test.

After the linear and phase-based motion processing, four videos that reflect the different modal shapes were reconstructed. The amplification factors in the linear and phase-based motion processing are illustrated in Table 5, and the screenshots of the four motion-magnified videos are shown in Figure 16. The magnification factors are restricted to avoid too many artifacts or blurs, and the motion of the different modal shapes can be better perceived from the video files. Because the motion corresponding to a specific mode cannot be separated simply by temporal filtering, the results of the original phase-based method are not presented here for predictable modal aliasing.

Table 5. Amplification factors in the Nanfeihe truss bridge case.

Mode Number	1	2	3	4
Linear (α_i)	10	15	25	30
Phase-based (β_i)	400	400	800	800

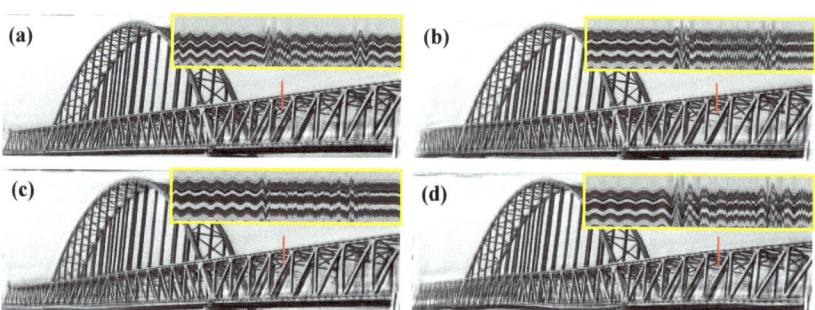

Figure 16. (a–d) Motion magnification results in the truss bridge test (8 orientations, quarter-octave bandwidth pyramids). More results are shown in the Supplementary videos.

When the train was driving across the bridge, a large deflection appeared on the bridge and then was attenuated by structural damping. For a simple single-DOF system, the transient vibration process is expressed as the combination of the damped vibration and the equal-amplitude vibration [35,36]. Due to the variation in the load, the attenuation of the deflection is in an unsteady state. Therefore, the differential of the decoupled temporal intensity variations is removed from the original data to investigate the influence of the damped vibration on the system. Figure 17 shows the results of the detected fourth variations (in Figure 15d). Several low-frequency peaks at 0.19, 0.34, 0.44, 0.58, and 0.73 Hz are found in the power spectrum. These peaks may be the resonant frequencies of the test bridge.

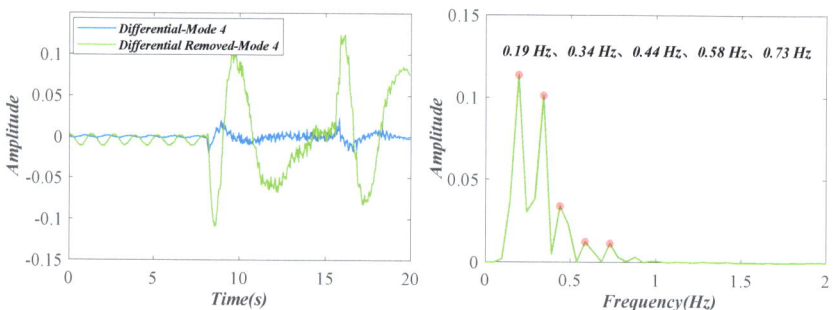

Figure 17. Damped vibration analysis result of the Nanfeihe truss bridge (the fourth variations).

4. Discussion

According to the theory of Eulerian video magnification, the spatial motion of a structure can be linearly approximated as temporal pixel variations. In hybrid motion processing, linear motion processing provides an effective approach to separate valuable temporal pixel variations and their corresponding spatial weights through an SVD and SCA. The FTP is utilized to improve the spatial weight matrices to achieve a better spatial consistency and noise reduction effect. Although the presented framework performs better on the vibration analysis than Eulerian linear processing, these two approaches have common limitations of a relatively low amplification factor and noise amplification.

Therefore, the output of linear motion processing is usually taken as the controllable input of the following phase-based motion processing. In practical applications, to minimize the residual noise, the motions generated in the pixel domain would be better to be just recognized enough by temporal phase variations, indicating that the factor α_i should not be too large. As spatial motion is generated into video, all temporal phase variations are usable, so temporal filtering can be omitted in phase-based motion processing. It is worth noting that for certain spatial motions, the phase amplification factor is still restricted by the spatial wavelength and the number of filters per octave for each orientation [14,15]. Therefore, the overall amplification factor in phase-based motion processing is not extended. As high-frequency components in images cannot be pushed as far as low-frequency components, breaking the restriction of the phase amplification factor will lead to artifacts or blurs. In the truss bridge test, blurs and artifacts are allowed to achieve a better perception of the modal shape.

Considering the fact that noises mainly exist in the high spatial-frequency part of the spatial weight, the FTP reserves a globally low spatial frequency of the spatial weights rather than directly reducing the amplification of these high spatial-frequency temporal variations [13,15]. In the current linear motion processing, the noise reduction process is simple and efficient in these controlled laboratory experiments without involving multi-scale decomposition. However, in the practical long-distance bridge test, the image quality is severely reduced due to the changing lighting and background conditions (e.g., clouds and the appearance of the high-speed train) [25]. This makes the FTP inefficient. The problem may be alleviated by setting masks for all video frames, but the whole process is too laborious, especially for high-speed videos. Moreover, in the field test, it is critical to remove the existing apparent motions in the video (Equation (10)) to ensure the stability of the modal shape visualization [20,25]. Based on the above analysis, we will attempt to address these issues in our future work and explore the practicability of the proposed framework for complex engineering structures.

5. Conclusions

In this paper, a hybrid motion processing framework that combines linear and phase-based motion processing is proposed, and its performance is evaluated through structural modal tests. By extracting, denoising, and manipulating the temporal intensity variations that are closely related to modal responses, the linear motion processing provides controllable, high-quality input for the subsequent phase-based motion processing, thus greatly improving the presentation of modal shapes. The proposed method is verified by two laboratory experiments on lightweight beams and a field test on a truss bridge. The experimental results indicate that the proposed motion processing framework can alleviate noise interference and obtain good results in subtle and long-distance motion observation. It should be pointed out that in the measurement of complex structures with a single camera, the motions in the image plane are considered as the projection of 3D vibration. Accurately representing global 3D motions of complex, large-scale engineering structures is challenging and significant. In addition to the issues listed in Discussions, we will further study the visualization of modal shapes in 3D space by extending the concept of motion amplification to 3D dynamic measurement techniques, such as multi-camera and structured light systems.

Supplementary Materials: The following are available online at https://www.mdpi.com/article/10.3390/s22239287/s1, Supplementary videos for Figure 8, Figure 12 and Figure 16.

Author Contributions: Conceptualization, methodology, software, writing—review and editing, D.Z.; investigation, experiments, data curation, writing—review and editing, A.Z.; data curation, experiments, W.H. and L.L.; supervision, Y.W. All authors have read and agreed to the published version of the manuscript.

Funding: This work was funded by the National Natural Science Foundation of China under Grant Nos. 51805006 and 51905005.

Institutional Review Board Statement: Not applicable.

Informed Consent Statement: Not applicable.

Data Availability Statement: Data sharing is not applicable to this article.

Acknowledgments: The authors would like to thank the editors and anonymous reviewers for their valuable comments and suggestions.

Conflicts of Interest: The authors declare no conflict of interest.

References

1. Fang, Z.; Yu, J.; Meng, X. Modal Parameters Identification of Bridge Structures from GNSS Data Using the Improved Empirical Wavelet Transform. *Remote Sens.* **2021**, *13*, 3375. [CrossRef]
2. Cakar, O.; Sanliturk, K.Y. Elimination of transducer mass loading effects from frequency response functions. *Mech. Syst. Signal Proc.* **2005**, *19*, 87–104. [CrossRef]
3. Zuo, D.; Hua, J.; Van Landuyt, D. A model of pedestrian-induced bridge vibration based on full-scale measurement. *Eng. Struct.* **2012**, *45*, 117–126. [CrossRef]
4. Olaszek, P.; Świercz, A.; Boscagli, F. The Integration of Two Interferometric Radars for Measuring Dynamic Displacement of Bridges. *Remote Sens.* **2021**, *13*, 3668. [CrossRef]
5. Khuc, T.; Catbas, F.N. Completely contactless structural health monitoring of real-life structures using cameras and computer vision. *Struct. Control. Health Monit.* **2017**, *24*, e1852. [CrossRef]
6. Feng, D.; Feng, M.Q. Computer vision for SHM of civil infrastructure: From dynamic response measurement to damage detection—A review. *Eng. Struct.* **2018**, *156*, 105–117. [CrossRef]
7. Kalybek, M.; Bocian, M.; Pakos, W.; Grosel, J.; Nikitas, N. Performance of Camera-Based Vibration Monitoring Systems in Input-Output Modal Identification Using Shaker Excitation. *Remote Sens.* **2021**, *13*, 3471. [CrossRef]
8. Seo, S.; Ko, Y.; Chung, M. Evaluation of Field Applicability of High-Speed 3D Digital Image Correlation for Shock Vibration Measurement in Underground Mining. *Remote Sens.* **2022**, *14*, 3133. [CrossRef]
9. Frankovský, P.; Delyová, I.; Sivák, P.; Bocko, J.; Živčák, J.; Kicko, M. Modal Analysis Using Digital Image Correlation Technique. *Materials* **2022**, *15*, 5658. [CrossRef] [PubMed]
10. Wang, Y.; Cai, J.; Zhang, D.; Chen, X.; Wang, Y. Nonlinear Correction for Fringe Projection Profilometry with Shifted-Phase Histogram Equalization. *IEEE Trans. Instrum. Meas.* **2022**, *71*, 1–9. [CrossRef]
11. Patil, K.; Srivastava, V.; Baqersad, J. A multi-view optical technique to obtain mode shapes of structures. *Measurement* **2018**, *122*, 358–367. [CrossRef]
12. Zhang, D.; Hou, W.; Guo, J.; Zhang, X. Efficient subpixel image registration algorithm for high precision visual vibrometry. *Measurement* **2021**, *173*, 108538. [CrossRef]
13. Wu, H.Y.; Rubinstein, M.; Shih, E.; Guttag, J.; Durand, F.; Freeman, W.T. Eulerian Video Magnification for Revealing Subtle Changes in the World. *ACM Trans. Graph.* **2012**, *31*, 1–8. [CrossRef]
14. Wadhwa, N.; Rubinstein, M.; Durand, F.; Freeman, W.T. Phase-Based Video Motion Processing. *ACM Trans. Graph.* **2013**, *32*, 1–10. [CrossRef]
15. Wadhwa, N.; Freeman, W.T.; Durand, F.; Wu, H.Y.; Guttag, J.V. Eulerian video magnification and analysis. *Commun. ACM* **2016**, *60*, 87–95. [CrossRef]
16. Chen, J.G.; Wadhwa, N.; Cha, Y.J.; Durand, F.; Freeman, W.T.; Buyukozturk, O. Modal identification of simple structures with high-speed video using motion magnification. *J. Sound Vibr.* **2015**, *345*, 58–71. [CrossRef]
17. Yang, Y.; Dorn, C.; Mancini, T.; Talken, Z.; Nagarajaiah, S.; Kenyon, G.; Farrar, C.; Mascareñas, D. Blind identification of full-field vibration modes of output-only structures from uniformly-sampled, possibly temporally-aliased (sub-Nyquist), video measurements. *J. Sound Vibr.* **2017**, *390*, 232–256. [CrossRef]
18. Yang, Y.; Dorn, C.; Mancini, T.; Talken, Z.; Kenyon, G.; Farrar, C.; Mascareñas, D. Blind identification of full-field vibration modes from video measurements with phase-based video motion magnification. *Mech. Syst. Signal Proc.* **2017**, *85*, 567–590. [CrossRef]
19. Davis, A.; Bouman, K.L.; Chen, J.G.; Rubinstein, M.; Büyüköztürk, O.; Durand, F.; Freeman, W.T. Visual Vibrometry: Estimating Material Properties from Small Motions in Video. *IEEE Trans. Pattern Anal. Mach. Intell.* **2017**, *39*, 732–745. [CrossRef]
20. Wadhwa, N.; Chen, J.G.; Sellon, J.B.; Wei, D.; Rubinstein, M.; Ghaffari, R.; Freeman, D.M.; Büyüköztürk, O.; Wang, P.; Sun, S.; et al. Motion microscopy for visualizing and quantifying small motions. *Proc. Natl. Acad. Sci. USA* **2017**, *114*, 11639–11644. [CrossRef]
21. Silva, M.; Martinez, B.; Figueiredo, E.; Costa, J.C.; Yang, Y.; Mascareñas, D. Nonnegative matrix factorization-based blind source separation for full-field and high-resolution modal identification from video. *J. Sound Vibr.* **2020**, *487*, 115586. [CrossRef]
22. Yang, Y.; Dorn, C.; Farrar, C.; Mascareñas, D. Blind, simultaneous identification of full-field vibration modes and large rigid-body motion of output-only structures from digital video measurements. *Eng. Struct.* **2020**, *207*, 110183. [CrossRef]
23. Eitner, M.; Miller, B.; Sirohi, J.; Tinney, C. Effect of broad-band phase-based motion magnification on modal parameter estimation. *Mech. Syst. Signal Proc.* **2021**, *146*, 106995. [CrossRef]
24. Siringoringo, D.M.; Wangchuk, S.; Fujino, Y. Noncontact operational modal analysis of light poles by vision-based motion-magnification method. *Eng. Struct.* **2021**, *244*, 112728. [CrossRef]

25. Chen, J.G.; Adams, T.M.; Sun, H.; Bell, E.S.; Oral, B. Camera-Based Vibration Measurement of the World War I Memorial Bridge in Portsmouth, New Hampshire. *J. Struct. Eng.* **2018**, *144*, 04018207. [CrossRef]
26. Zhang, D.; Guo, J.; Lei, X.; Zhu, C. Note: Sound recovery from video using SVD-based information extraction. *Rev. Sci. Instrum.* **2016**, *87*, 198–516. [CrossRef] [PubMed]
27. Qin, S.; Zhu, C.; Jin, Y. Sparse Component Analysis Based on Hierarchical Hough Transform. *Circuits Syst. Signal Process.* **2017**, *36*, 1569–1585.
28. Xu, Y.; Brownjohn, J.M.; Hester, D. Enhanced sparse component analysis for operational modal identification of real-life bridge structures. *Mech. Syst. Signal Proc.* **2019**, *116*, 585–605. [CrossRef]
29. Takeda, M.; Mutoh, K. Fourier transform profilometry for the automatic measurement of 3-D object shapes. *Appl. Optics* **1983**, *22*, 3977. [CrossRef]
30. Berryman, F.; Pynsent, P.; Cubillo, J. A theoretical comparison of three fringe analysis methods for determining the three-dimensional shape of an object in the presence of noise. *Opt. Lasers Eng.* **2003**, *39*, 35–50. [CrossRef]
31. Mittal, A.; Moorthy, A.K.; Bovik, A.C. No-Reference Image Quality Assessment in the Spatial Domain. *IEEE Trans. Image Process.* **2012**, *21*, 4695–4708. [CrossRef]
32. Feng, M.Q.; Fukuda, Y.; Feng, D.; Mizuta, M. Nontarget Vision Sensor for Remote Measurement of Bridge Dynamic Response. *J. Bridge Eng.* **2015**, *20*, 04015023.1–04015023.12. [CrossRef]
33. Hermanns, L.; Gimenez, J.G.; Alarcon, E. Efficient computation of the pressures developed during high-speed train passing events. *Comput. Struct.* **2005**, *83*, 793–803. [CrossRef]
34. Feng, D.; Feng, M.Q. Model Updating of Railway Bridge Using In Situ Dynamic Displacement Measurement under Trainloads. *J. Bridge Eng.* **2015**, *20*, 04015019. [CrossRef]
35. Xiao, X.; Zhang, Y.; Shen, W.; Kong, F. A stochastic analysis method of transient responses using harmonic wavelets, Part 1:Time-invariant structural systems. *Mech. Syst. Signal Proc.* **2021**, *160*, 107870. [CrossRef]
36. Xiao, X.; Zhang, Y.; Shen, W. A stochastic analysis method of transient responses using harmonic wavelets, part 2: Time-dependent vehicle-bridge systems. *Mech. Syst. Signal Proc.* **2022**, *162*, 107871. [CrossRef]

Article

Low-Complexity Lossless Coding of Asynchronous Event Sequences for Low-Power Chip Integration

Ionut Schiopu * and Radu Ciprian Bilcu

Tampere Handset Camera Innovation Lab, Huawei Technologies Oy (Finland) Co., Ltd., 33720 Tampere, Finland
* Correspondence: ionut.schiopu@huawei.com

Citation: Schiopu, I.; Bilcu, R.C. Low-Complexity Lossless Coding of Asynchronous Event Sequences for Low-Power Chip Integration. *Sensors* 2022, 22, 10014. https://doi.org/10.3390/s222410014

Academic Editors: Ittetsu Taniguchi, Jinjia Zhou and Xin Jin

Received: 21 November 2022
Accepted: 16 December 2022
Published: 19 December 2022

Publisher's Note: MDPI stays neutral with regard to jurisdictional claims in published maps and institutional affiliations.

Copyright: © 2022 by the authors. Licensee MDPI, Basel, Switzerland. This article is an open access article distributed under the terms and conditions of the Creative Commons Attribution (CC BY) license (https://creativecommons.org/licenses/by/4.0/).

Abstract: The event sensor provides high temporal resolution and generates large amounts of raw event data. Efficient low-complexity coding solutions are required for integration into low-power event-processing chips with limited memory. In this paper, a novel lossless compression method is proposed for encoding the event data represented as asynchronous event sequences. The proposed method employs only low-complexity coding techniques so that it is suitable for hardware implementation into low-power event-processing chips. A first, novel, contribution consists of a low-complexity coding scheme which uses a decision tree to reduce the representation range of the residual error. The decision tree is formed by using a triplet threshold parameter which divides the input data range into several coding ranges arranged at concentric distances from an initial prediction, so that the residual error of the true value information is represented by using a reduced number of bits. Another novel contribution consists of an improved representation, which divides the input sequence into same-timestamp subsequences, wherein each subsequence collects the same timestamp events in ascending order of the largest dimension of the event spatial information. The proposed same-timestamp representation replaces the event timestamp information with the same-timestamp subsequence length and encodes it together with the event spatial and polarity information into a different bitstream. Another novel contribution is the random access to any time window by using additional header information. The experimental evaluation on a highly variable event density dataset demonstrates that the proposed low-complexity lossless coding method provides an average improvement of 5.49%, 11.45%, and 35.57% compared with the state-of-the-art performance-oriented lossless data compression codecs Bzip2, LZMA, and ZLIB, respectively. To our knowledge, the paper proposes the first low-complexity lossless compression method for encoding asynchronous event sequences that are suitable for hardware implementation into low-power chips.

Keywords: low-power electronics; low-complexity codec; lossless compressio; event camera

1. Introduction

The recent research breakthroughs in the neuromorphic engineering domain have made possible the development of a new type of sensor, called the event camera, which is bioinspired by the human brain, as each pixel operates individually and mimics the behaviour of a separate nerve cell. In contrast to the conventional camera, in which all pixels are designed to capture the intensity of the incoming light at the same time, the event camera sensor reports only the changes of the incoming light intensity above a threshold, at any timestamp, and at any pixel position by triggering a sequence of asynchronous events (sometimes called spikes); otherwise it remains silent. Because each pixel detects and reports independently only the change in brightness, the event camera sensor proposes a new paradigm shift for capturing visual data.

The event camera provides a series of important technological advantages, such as a high temporal resolution as the asynchronous events can be triggered at a minimum timestamp distance of only $1\,\mu s$ ($10^{-6}\,s$), i.e., the event sensor can achieve a frame rate of up to 1 million (M) frames per second (fps). This is made possible thanks to the remarkable novel event camera feature of capturing all dynamic information without unnecessary static information (e.g., background), which is an extremely useful feature for capturing high-speed motion scenes for which the conventional camera usually fails to provide a good performance. Two types of sensors are currently available on the market: (i) the dynamic vision sensor (DVS) [1], which captures only the event modality; and (ii) the dynamic and active-pixel vision sensor (DAVIS) [2], which is comprised of a DVS sensor and an active pixel sensor (APS), i.e., it captures a sequence of conventional camera frames and their corresponding event data. The event camera sensors are now widely used in the computer vision domain, wherein the RGB and event-based solutions already provide an improved performance compared with state-of-the-art RGB-based solutions for applications such as deblurring [3], feature detection and tracking [4,5], optic flow estimation [6], 3D estimation [7], superresolution [8], interpolation [9], visual odometry [10], and many others. For more details regarding event-based applications in computer vision, please see the comprehensive literature review presented in [11]. To achieve high frame rates, the captured asynchronous event sequences reach high bit-rate levels when stored using the raw event representation of 8 bytes (B) per event provided by the event camera. Therefore, for better preprocessing of event data on low-power event-processing chips, novel low-complexity and efficient event coding solutions are required to be able to store without any information loss the acquired raw event data. In this paper, a novel low-complexity lossless compression method is proposed for efficient-memory representation of the asynchronous event sequences by employing a novel low-complexity coding scheme so that the proposed codec is suitable for hardware implementation into low-cost event signal processing (ESP) chips.

The event data compression domain is understudied whereas the sensor's popularity continues to grow thanks to improved technical specifications offered by the latest class of event sensors. The problem was tackled in only a few articles that propose to either encode the raw asynchronous event sequences generated by the sensor with or without any information loss [12–14], or to first preprocess the event data from a sequence of synchronous event frames (EFs) that are finally encoded by employing a video coding standard [15,16]. The EF sequences are formed by using an event-accumulation process that consists of splitting the asynchronous event sequence into spatiotemporal neighbourhoods of time intervals, processing the events triggered in a single time interval, and then generating a single event for each pixel position in the EF. These performance-oriented coding solutions are too complex for hardware implementation in the ESP chip designed with limited memory, and may be integrated only in a system on a chip (SoC) wherein enough computation power and memory is available.

In our prior work [17,18], we proposed employing an event-accumulation process which first splits each asynchronous event sequence into spatiotemporal neighbourhoods by using different time-window values, and then generates the EF sequence by using a sum-accumulation process, whereby the events triggered in a time window are represented by a single event that is set as the sign of the event polarity sum and stored at the corresponding pixel position. In [17], we proposed a performance-oriented, context-based lossless image codec for encoding the sequence of event camera frames, in which the event spatial information and the event polarity are encoded separately by using the event map image (EMI) and the concatenated polarity vector (CPV). One can note that the lossless compression codec proposed in [17] is suitable for hardware implementation in SoC chips. In [18], we proposed a low-complexity lossless coding framework for encoding event camera frames by adapting the run-length encoding scheme and Elias coding [19] for EF coding. One can note that the low-complexity lossless compression codec proposed in [18] is suitable for hardware implementation in ESP chips. The goal of this work is to propose a novel low complexity-oriented lossless compression codec for encoding asynchronous event sequences, suitable for hardware implementation in ESP chips.

In summary, the novel contributions of this work are summarized as follows.

(1) A novel low-complexity lossless compression method for encoding raw event data represented as asynchronous event sequences, which is suitable for hardware implementation into ESP chips.
(2) A novel low-complexity coding scheme for encoding residual errors by dividing the input range into several coding ranges arranged at concentric distances from an initial prediction.
(3) A novel event sequence representation that removes the event timestamp information by dividing the input sequence into ordered same-timestamp event subsequences that can be encoded in separated bit streams.
(4) A lossless event data codec that provides random access (RA) to any time window by using additional header information.

The remainder of this paper is organized as follows. Section 2 presents an overview of state-of-the-art methods. Section 3 describes the proposed low-complexity lossless coding framework. Section 4 presents the experimental evaluation of the proposed codecs. Section 5 draws the conclusions of this work.

2. State-of-the-Art Methods

To achieve an efficient representation of the large amount of event data, a first approach was proposed to losslessly (without any information loss) encode the asynchronous event representation. In [12], a lossless compression method is proposed by removing the redundancy of the spatial and temporal information by using three strategies: adaptive macrocube partitioning structure, the address-prior mode, and the time-prior mode. The method was extended in [13] by introducing an event sequence octree-based cube partition and a flexible intercube prediction method based on motion estimation and motion compensation. However, the coding performance of these methods (based on the spike coding strategy) remains limited.

In another approach, the asynchronous event representation is compressed by employing traditional lossless data compression methods. In [14], the authors present a coding performance comparison study of different traditionally based lossless data compression strategies when employed to encode raw event data. The study shows that traditional dictionary-based methods for data compression provide the best performance. The dictionary-based approach consists of searching for matches of data between the data to be compressed and a set of strings stored as a dictionary, in which the goal is to find the best match between the information maintained in the dictionary and the data to be compressed. One of the most well-known algorithms for lossless data compression is the Lempel-Ziv 77 (LZ77) algorithm [20], which was created by Lempel and Ziv in 1977. LZ77 iterates sequentially through the input string and stores any new match into a search buffer. The Zeta Library (ZLIB) [21], an LZ77 variant called deflation, proposed a strategy whereby the input data is divided into a sequence of blocks. The Lempel–Ziv–Markov chain algorithm (LZMA) [22] is an advanced dictionary-based codec developed by Igor Pavlov for lossless data compression, which was first used in the 7-Zip open source code. The Bzip2 algorithm is based on the well-known Burrows–Wheeler transform [23] for block sorting, which operates by applying a reversible transformation to a block of input data.

In a more recent approach [24], the authors propose to treat the asynchronous event sequence as a point cloud representation and to employ a lossless compression method based on a point cloud compression strategy. One can note that the coding performance of such a method depends on the performance of the geometry-based point cloud compression (G-PCC) algorithm used in the algorithm design.

Many of the upper-level applications prefer to consume the event data as an "intensity-like" image rather than asynchronous events sequence, wherein several event-accumulation processes are proposed [25–30] to form the EF sequence. Hence, in another approach, several methods are proposed to losslessly encode the generated EF sequence. The study in [14] was extended in [15] by proposing a time aggregation-based lossless video encoding method based on the strategy of accumulating events over a time interval by creating two event frames that count the number positive and negative polarity events, which are concatenated and encoded by the high-efficiency video coding (HEVC) standard [31]. Similarly, the coding performance depends on the performance of the video coding standard employed to encode the concatenated frames.

To further improve event data representation, another approach was proposed to encode the asynchronous event sequences by relaxing the lossless compression constraint problem and accepting information loss. In [32], the authors propose a macrocuboids partition of the raw event data, and they employ a novel spike coding framework, inspired by video coding, to encode spike segments. In [16], the authors propose a lossy coding method based on a quad-tree segmentation map derived from the adjacent intensity images. One can note that the information loss introduced by such methods might affect the performance of the upper-level applications.

3. Proposed Low-Complexity Lossless Coding Framework

Let us consider an event camera having a $W \times H$ pixel resolution. Any change of the incoming light intensity triggers an asynchronous event, $e_i = (x_i, y_i, p_i, t_i)$, which stores (based on the sensors representation) the following information in 8 B of memory:

- spatial information $(x_i, y_i), \forall x_i \in [1, H], y_i \in [1, W]$, i.e., the pixel positions where the event was triggered;
- polarity information $p_i \in \{-1, 1\}$, where the symbol "-1" signals a decrease and symbol "1" signals an increase in the light intensity; and
- timestamp t_i, the time when the event was triggered.

Hence, an asynchronous event sequence, denoted as $\mathcal{S}_\mathcal{T} = \{e_i\}_{i=1,2,\ldots,N_e}$, collects N_e events triggered over a time period of $\mathcal{T}$ µs. The goal of this paper is to encode $\mathcal{S}_\mathcal{T}$ by employing a novel, low-complexity lossless compression algorithm.

Figure 1 depicts the proposed low-complexity lossless coding framework scheme for encoding asynchronous event sequences. A novel sequence representation groups the same-timestamp events in subsequences and reorders them. Each same-timestamp subsequence is encoded in turn by the proposed method, called low-complexity lossless compression of asynchronous event sequences (LLC-ARES). LLC-ARES is built based on a novel coding scheme, called the triple threshold-based range partition (TTP).

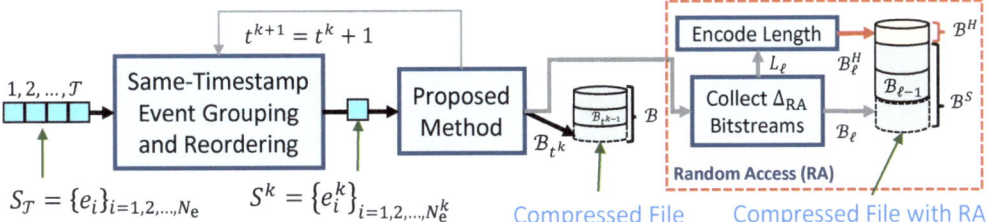

Figure 1. The proposed low-complexity lossless coding framework. The input asynchronous event sequence, $\mathcal{S}_\mathcal{T}$, is first represented by using the proposed event representation as a set of same-timestamp subsequences, $\mathcal{S}^k$, having same-timestamp t^k, and then encoded losslessly by employing the proposed method. The output bitstream of each same-timestamp subsequence can be stored in memory as a compressed file. Moreover, it can also be collected as a package bitstream for all the timestamps found in a time period Δ_{RA} and then stored in memory together with bitstream-length information stored as a header as a compressed file with RA, so that the proposed codec can provide RA to any time window of size Δ_{RA}.

Section 3.1 presents the proposed sequence representation. Section 3.2 presents the proposed low-complexity coding scheme. Section 3.3 presents the proposed method.

3.1. Proposed Sequence Representation

An input asynchronous event sequence, $\mathcal{S}_\mathcal{T}$, is arranged as a set of same-timestamp subsequences, $\mathcal{S}_\mathcal{T} = \{S^k\}_{k=0,1,\dots,\mathcal{T}-1}$, where each same-timestamp subsequence $S^k = \{e_i^k\}_{i=1,2,\dots,N_e^k} = \{(x_i^k, y_i^k, p_i^k)\}_{i=1,2,\dots,N_e^k}$ collects all N_e^k events in $\mathcal{S}_\mathcal{T}$ triggered at the same timestamp t^k. One can note that at the decoder side the timestamp information is recovered based on the subsequence length information, $\{N_e^k\}_{k=0,1,\dots,\mathcal{T}-1}$, i.e., $t^k = k$ is set to all N_e^k events. Each S^k is ordered in the ascending order of the largest spatial information dimension, e.g., $y_i^k < y_{i+1}^k$. However, if $y_i^k = y_{i+1}^k$, then S^k is further ordered in the ascending order of the remaining dimension, i.e., $x_i^k < x_{i+1}^k$.

Figure 2 depicts the proposed sequence representation and highlights the difference between the sensor's event-by-event (EE) order, depicted on the left side, and the same-timestamp (ST) order, depicted on the right side. Note that the EE order proposes to write to file, in turn, each event e_i. Although the proposed ST order proposes to write to file the number of events of each same-timestamp subsequence, N_e^k having the same-timestamp t^k, and, if $N_e^k > 0$, it is followed by the spatial and the event information of all same-timestamp events, i.e., $\{x_i\}_{i=1:N_e^k}, \{y_i\}_{i=1:N_e^k}, \{p_i\}_{i=1:N_e^k}$. Section 4 shows that the state-of-the-art dictionary-based data compression methods provide an improved performance when the proposed ST order is employed to represent the input data compared with the EE order.

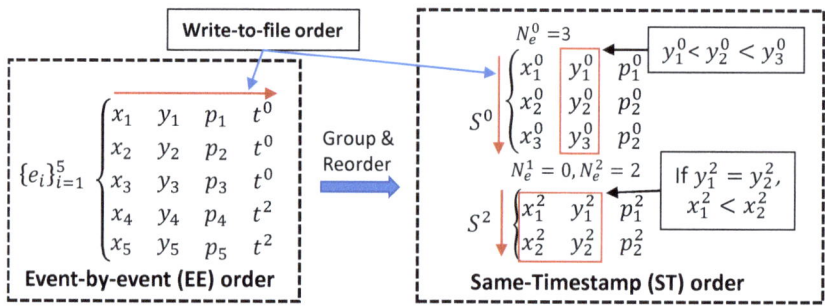

Figure 2. The proposed representation based the proposed same-timestamp (ST) order (on the right) in comparison with the sensor's event-by-event (EE) order (on the left). The red arrow shows the write-to-file order used to generate the input data files feed to the traditional methods.

3.2. Proposed Triple Threshold-Based Range Partition (TTP)

For hardware implementation of the proposed event data codec into low-power event-processing chips, a novel low-complexity coding scheme is proposed. The binary representation range of the residual error is partitioned into smaller intervals selected by using a short-depth decision tree designed based on a triple threshold, $\Delta = (\delta_1, \delta_2, \delta_3)$. Hence, the input range is partitioned into several smaller coding ranges arranged at concentric distances from the initial prediction.

Let us consider the case of encoding $x \in [1, H]$, i.e., a finite range, by using the prediction $\hat{x}$ by writing the binary representation of the residual error $\epsilon = x - \hat{x}$ on exactly n_ϵ bits. Because on the decoder side n_ϵ is unknown, the triple threshold Δ is used to create a decision tree having the role of partitioning the input range $[1, H]$ into five types of coding ranges (see Figure 3a), where either the binary representation of ϵ is represented by using a different number of bits or the binary representation of x is written by using a different number of bits.

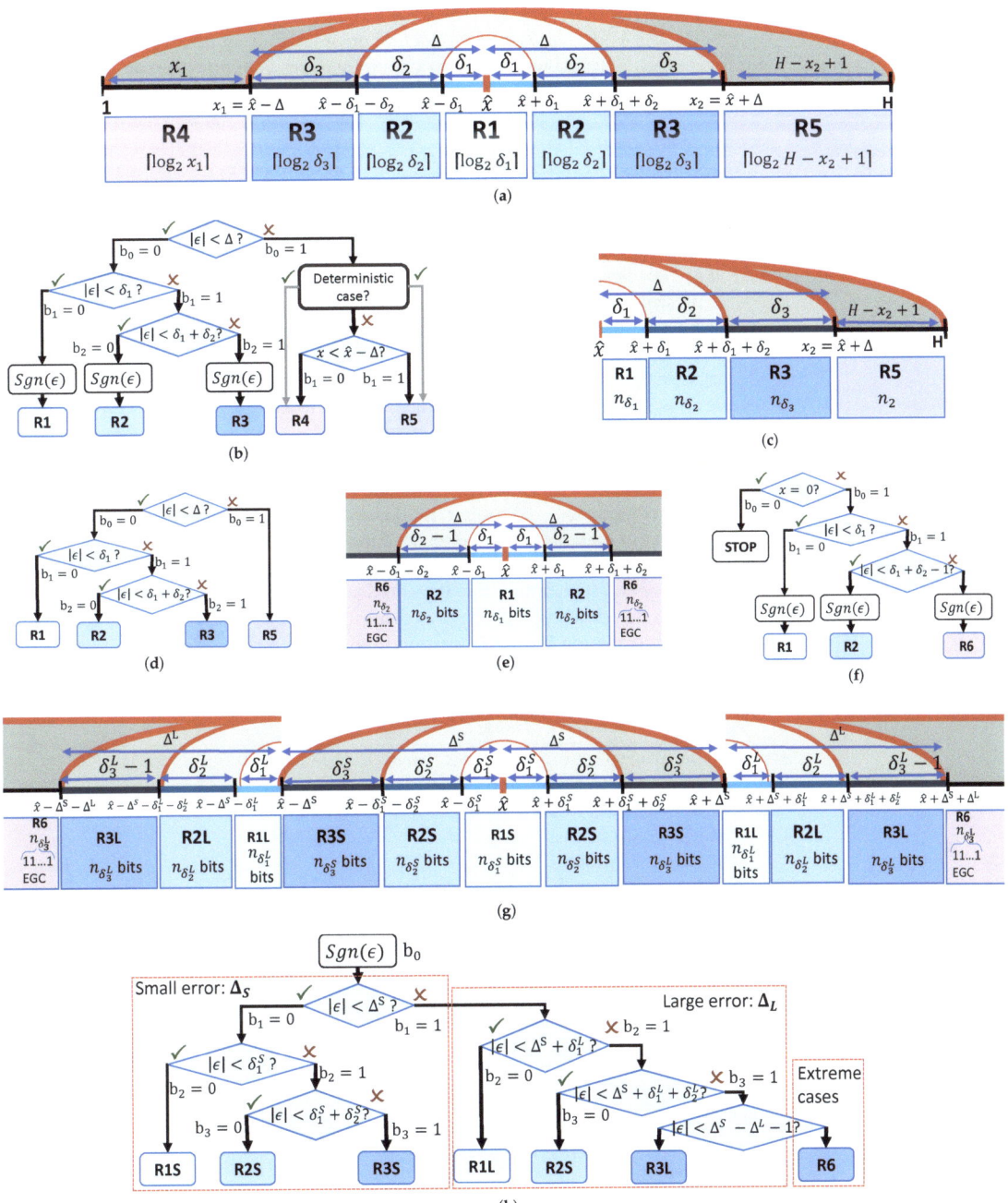

Figure 3. The proposed low-complexity coding scheme, triple threshold-based range partition (TTP). (**a**) TTP range partition. (**b**) TTP decision tree. (**c**) TTP$_y$ range partition. (**d**) TTP$_y$ decision tree. (**e**) TTP$_e$ range partition. (**f**) TTP$_e$ decision tree. (**g**) TTP$_L$ range partition. (**h**) TTP$_L$ range partition.

Let us denote $\Delta = \delta_1 + \delta_2 + \delta_3, x_1 = \hat{x} - \Delta, x_2 = \hat{x} + \Delta, n_{\delta_j} = \lceil \log_2 \delta_j \rceil, \forall j = 1,2,3$, $n_1 = \lceil \log_2 x_1 \rceil$, and $n_2 = \lceil \log_2(H - x_2 - 1) \rceil$. The 1st range, R1, is defined by using δ_1 as $(\hat{x} - \delta_1, \hat{x} + \delta_1)$ to represent any residual error $|\epsilon| < \delta_1$ on n_{δ_1} bits plus an additional bit for $sign(\epsilon)$. The 2nd range, R2, is defined by using δ_2 to represent any residual error $|\epsilon| - \delta_1 < \delta_2$ on n_{δ_2} bits plus a sign bit, i.e., $x \in (\hat{x} - \delta_1 - \delta_2, \hat{x} - \delta_1]$ for $\epsilon < 0$ and $x \in [\hat{x} + \delta_1, \hat{x} + \delta_1 + \delta_2)$ for $\epsilon \geq 0$. Similarly, the 3rd range, R3, is defined by using δ_3 to represent any residual error $|\epsilon| - \delta_1 - \delta_2 < \delta_3$ on n_{δ_3} bits plus a sign bit. The 4th (R4) and 5th (R5) ranges are defined for $|\epsilon| \geq \Delta$ and used to represent $x - 1$ on n_1 bits and $H - x$ on n_2 bits, respectively.

Figure 3b depicts the decision tree defined by checking the following four constraints:

(c1) b_0 is set by checking $|\epsilon| < \Delta$. If true then $b_0 = 0$; otherwise, $b_0 = 1$.
(c2) If $b_0 = 0$, then b_1 is set by checking $|\epsilon| < \delta_1$. If true, then $b_1 = 0$ and R1 is employed to represent ϵ on $n_\epsilon = n_{\delta_1} + 1$ bits; otherwise $b_1 = 1$.
(c3) If $b_1 = 1$, then b_2 is set by checking $|\epsilon| < \delta_1 + \delta_2$. If true then $b_2 = 0$ and R2 is employed to represent ϵ on $n_\epsilon = n_{\delta_2} + 1$ bits. Otherwise, $b_1 = 1$ and R3 is used to represent ϵ on $n_\epsilon = n_{\delta_3} + 1$ bits.
(c4) If $b_0 = 1$, then b_1 is set by checking $x \leq x_1$. If true, then $b_1 = 0$ and R4 is employed to represent $x - 1$ on n_1 bits. Otherwise, $b_1 = 1$ and R5 is used to represent $H - x$ on n_2 bits.

Note that the range $[1, x_1]$ contains x_1 possible values. To fully utilize the entire set of code words (i.e., including $00 \cdots 0$ having n_1 bits length), $x - 1$ is represented on n_1 bits.

Algorithm 1 presents the pseudocode of the basic implementation of the TTP encoding algorithm. It is employed to represent a general value x by using the prediction $\hat{x}$, the support range $[1, H]$, and the triple threshold parameter, Δ, as output bitstream B, which contains the decision tree bits, followed by the binary representation of the required additional information for the corresponding coding range. Algorithm 2 presents the pseudocode of the basic implementation of the corresponding TTP decoding algorithm.

Algorithm 1: Encode a general x by using TTP

Data: True value x, prediction $\hat{x}$, range $[1, H]$, and triple threshold Δ;
Result: Output bitstream B;

1 $B(0) \leftarrow 0; B(1) \leftarrow 0; \epsilon = x - \hat{x}; \Delta = \delta_1 + \delta_2 + \delta_3$;
2 **if** $|\epsilon| < \Delta$ **then**
3   **if** $|\epsilon| < \delta_1$ **then**   // R1 Range
4        $B(2 : \lceil \log_2 \delta_1 \rceil + 2) \leftarrow [sign(\epsilon);$ Write $|\epsilon|$ on $\lceil \log_2 \delta_1 \rceil$ bits$]$;
5   **else**
6   **if** $|\epsilon| < \delta_1 + \delta_2$ **then**   // R2 Range
7           $B(1 : \lceil \log_2 \delta_2 \rceil + 3) \leftarrow [1; 0; sign(\epsilon);$ Write $|\epsilon| - \delta_1$ on $\lceil \log_2 \delta_2 \rceil$ bits$]$;
8   **else**   // R3 Range
9           $B(1 : \lceil \log_2 \delta_3 \rceil + 3) \leftarrow [1; 1; sign(\epsilon);$ Write $|\epsilon| - \delta_1 - \delta_2$ on $\lceil \log_2 \delta_3 \rceil$ b$]$;
10 **else**
11   **if** $x \leq \hat{x} - \Delta$, **then**   // R4 Range
12       $B(2 : \lceil \log_2(\hat{x} - \Delta) \rceil + 1) \leftarrow [$Write $x - 1$ on $\lceil \log_2(\hat{x} - \Delta) \rceil$ bits$]$;
13   **else**   // R5 Range
14       $B(1 : \lceil \log_2(H - \hat{x} - \Delta - 1) \rceil + 1) \leftarrow [1; H - x$ on $\lceil \log_2(H - \hat{x} - \Delta - 1) \rceil]$;
15 **Return** B;

Algorithm 2: Decode a general x by using TTP

Data: Bitstream B; prediction $\hat{x}$, range $[1, H]$, and triple threshold Δ;
Result: True value x;

1. **if** $B(0) = 0$ **then**
2.   **if** $B(1) = 0$ **then**   // R1 Range
3. | $sign_\epsilon \leftarrow B(2); \epsilon_{abs} \leftarrow Dec2bin(B(3:\lceil \log_2 \delta_1 \rceil + 2))$;
4.   **else**
5.   **if** $B(2) = 0$ **then**   // R2 Range
6. | $sign_\epsilon \leftarrow B(3); \epsilon_{abs} \leftarrow \delta_1 + Dec2bin(B(4:\lceil \log_2 \delta_2 \rceil + 3))$;
7.   **else**   // R3 Range
8. | $sign_\epsilon \leftarrow B(3); \epsilon_{abs} \leftarrow \delta_1 + \delta_2 + Dec2bin(B(4:\lceil \log_2 \delta_3 \rceil + 3))$;
9. $x \leftarrow sign_\epsilon \cdot \epsilon_{abs}$;
10. **else**
11.   **if** $B(1) = 0$ **then**   // R4 Range
12. | $x \leftarrow 1 + Dec2bin(B(2:\lceil \log_2(\hat{x} - \Delta) \rceil + 2))$;
13.   **else**   // R5 Range
14. | $x \leftarrow H - Dec2bin(B(2:\lceil \log_2(H - \hat{x} - \Delta - 1) \rceil + 2))$;
15. **Return** x;

Section 3.2.1 presents the deterministic cases that may occur. Section 3.2.2 analyses the different algorithmic variations proposed to encode the data structures in the proposed event representation that have different properties.

3.2.1. Deterministic Cases

In some special cases, some part of the information can be directly determined from the current coding context. For example, if x_1 or x_2 is outside the finite range (see Figure 4a), then R4 or R5 does not exist and the context tree is built without checking condition (c4), i.e., in such case one bit is saved. More exactly, steps 11–14 in Algorithms 1 and 2 are replaced with either step 12 (encode/decode using R4) or step 14 (encode/decode using R5).

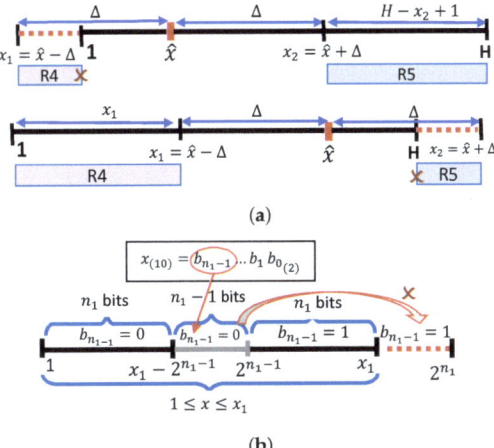

Figure 4. Deterministic cases: (**a**) if $x_1 < 1$ or $x_2 > H$, then condition (c4) is not checked when building the context tree and one bit is saved. (**b**) If $x \in (x_1 - 2^{n_1-1}, 2^{n_1-1}]$, then x is represented by using one bit less than in the case when $x \in [1, x_1 - 2^{n_1-1}]$ or $x \in (2^{n_1-1}, x_1]$.

Moreover, because x_1 and $x_2' = H - x_2 + 1$ are not power-2 numbers, the most significant bit of x, b_{n_1-1}, is 0, thanks to the constraint $1 \leq x \leq x_1$ and $1 \leq x \leq x_2'$, respectively. Figure 4b shows that if $x \in (x_1 - 2^{n_1-1}, 2^{n_1-1}]$ and b_{n_1-1} would be set as

1, then $x > x_1$ and the constraint would be violated. Hence, b_{n_1-1} is always set 0 if $x \in (x_1 - 2^{n_1-1}, 2^{n_1-1}]$, (or similarly when $x \in (x'_2 - 2^{n'_2-1}, 2^{n'_2-1}]$).

3.2.2. Algorithm Variations

The basic implementation of the TTP algorithm was modified for encoding different types of data. Let us denote $\epsilon_{x_i^k} = x_i^k - \hat{x}_i^k$ and $\epsilon_{y_i^k} = y_i^k - \hat{y}_i^k$. Then the sequence $\{x_i^k\}_{i=1,2,\ldots,N_e^k}$ is encoded by using version TTP$_x$, where $\epsilon_{y_i^k}$ is used to detect another deterministic case: if $\epsilon_{y_i^k} = 0$, then $\hat{x}_i^k = x_{i-1}^k$ and the sign bit is saved (see Figure 2 (ST order)). The sequence $\{y_i^k\}_{i=2,3,\ldots,N_e^k}$ having $\epsilon_{y_i^k} \geq 0$ (thanks to ST order) is encoded by using version TTP$_y$, which is designed to encode a general value x found in range $[\hat{x}, H]$. Figure 3c,d show the TTP$_y$ range partitioning and decision tree, respectively.

Some data types have a very large or infinite support range. The sequence of number of events of each timestamp, $\{N_e^k\}_{k=0,1,\ldots,\mathcal{T}-1}$, is encoded by using version TTP$_e$. Note that $N_e^k \in [0, HW]$; however, there is a very low probability of having a large majority of pixels triggered with the same timestamp. Therefore, because N_e is usually very small, TTP$_e$ is designed to use the doublet threshold $\Delta_e = (\delta_1, \delta_2)$, as experiments show that a triplet threshold does not improve the coding performance. Figure 3e shows the TTP$_e$ range partitioning, where the values $0, 1, \ldots, \delta_2 - 2$ are encoded by R2 as the last value, $\delta_2 - 1$ (having the binary representation as n_{δ_2} bits of 1, i.e., $\underbrace{11\ldots1}_{n_{\delta_2}}$), signals the use of R6 to encode $|\epsilon| - \Delta - 2$ by using a simple coding technique, the Elias gamma coding (EGC) [19]. Figure 3f shows the decision tree, where $N_e^k = 0$ (i.e., $S^k = \emptyset$) is encoded by the first bit of the decision tree.

Finally, TTP$_L$ is designed to encode the length of the package bitstream $\mathcal{B}_\ell$, denoted as L_ℓ (see Section 3.3.3). TTP$_L$ defines seven partition intervals by using two triple thresholds: $\Delta^S = (\delta_1^S, \delta_2^S, \delta_3^S)$ is used for encoding small errors using R1S, R2S, and R3S, and $\Delta^L = (\delta_1^L, \delta_2^L, \delta_3^L)$ is used for encoding large errors using R1L, R2L, and R3L. Similar to TTP$_e$, R6 is signalled in R3L by using the last value $\delta_3^L - 1$ and $|\epsilon| - \Delta^S - \Delta^L - 2$ is encoded by employing EGC [19].

3.3. Proposed Method

The proposed method, LLC-ARES, employs the proposed representation to generate the set of same-timestamp subsequences, $\{S^k\}_{k=0,1,\ldots,\mathcal{T}-1}$ (see Section 3.1). Subsequence S^k is encoded as bitstream $\mathcal{B}_{t^k}$ by using Algorithm 3, which employs the proposed coding scheme, TTP (see Section 3.2). The compressed file collects these bitstreams as $\mathcal{B} = [\mathcal{B}_{t^0} \mathcal{B}_{t^1} \cdots \mathcal{B}_{t^{\mathcal{T}-1}}]$.

Algorithm 3: Encode the subsequence of ordered events

Data: $S_k = \{(x_i^k, y_i^k, p_i^k)\}_{i=1:N_e^k}$ of same timestamp t^k, $\{N_e^j\}_{j=k-3:k-1}$, H, W;
Result: Output bitstream B_{t^k};

1. $\hat{N}_e^k \leftarrow$ **Predict** N_e^k using $\{N_e^j\}_{j=k-3:k-1}$;
2. $B_{t^k} \leftarrow$ **Encode** N_e^k using $\text{TTP}_e(\hat{N}_e^k, \Delta_e)$;
3. **if** $N_e^k > 0$ **then**
4. $\quad B_{t^k} \leftarrow$ **Encode** y_1^k using $\text{TTP}_x(\hat{y}_r^k, \epsilon_{y_1^k} > 0, [1, W], \Delta^{e1})$;
5. $\quad B_{t^k} \leftarrow$ **Encode** x_1^k using $\text{TTP}_x(\hat{x}_r^k, \epsilon_{y_1^k} > 0, [1, H], \Delta^{e1})$;
6. $\quad B_{t^k} \leftarrow$ **Encode** p_1^k as 0 for $p_1^k = -1$ and 1 for $p_1^k = 1$;
7. $\quad$ **for** $i = 2, 3, \ldots, N_e^k$ **do**
8. $\quad\quad B_{t^k} \leftarrow$ **Encode** y_i^k using $\text{TTP}_y(y_{i-1}^k, [\hat{y}_i^k, W], \Delta_W^k)$;
9. $\quad\quad \hat{x}_i^k \leftarrow$ **Predict** x_i^k using $\{x_j^k\}_{j=1,2,\ldots,i-1}$;
10. $\quad\quad B_{t^k} \leftarrow$ **Encode** x_i^k using $\text{TTP}_x(\hat{x}_i^k, \epsilon_{y_i^k}, [1, H], \Delta_H^k)$;
11. $\quad\quad B_{t^k} \leftarrow$ **Encode** p_i^k as 0 for $p_i^k = -1$ and 1 for $p_i^k = 1$;
12. $\quad$ **end**
13. $\quad \Delta_H^{k+1} \leftarrow$ **Update** Δ_H^k using $\epsilon^k = y_{N_e^k}^k - y_1^k$;
14. $\quad \Delta_W^{k+1} \leftarrow$ **Update** Δ_W^k using $\epsilon^k = y_{N_e^k}^k - y_1^k$;
15. **end**
16. **Return** B_{t^k};

Algorithm 3 encodes the following data structures:

(i) Encode N_e^k by employing TTP_e using $\hat{N}_e^k$, computed by (1), and Δ_e;
(ii) Encode e_1^k as follows:
   (ii.1) y_1^k by employing TTP_x using $\hat{y}_r^k$ computed by (2), range $[1, W]$, and Δ^{e1};
   (ii.2) x_1^k by employing TTP_x using $\hat{x}_r^k$ computed by (2), range $[1, H]$, and Δ^{e1}; and
   (ii.3) p_1^k using binarization;
(iii) The remaining events are encoded as follows:
   (iii.1) y_i^k by employing TTP_y using $\hat{y}_i^k = y_{i-1}^k$, range $[\hat{y}_i^k, W]$, and Δ_W^k;
   (iii.2) x_i^k by employing TTP_x using $\hat{x}_i^k$ computed by (3), $\epsilon_{y_i^k}$, range $[1, H]$, and Δ_H^k; and
   (iii.3) p_i^k using binarization.
(iv) Update the triple thresholds Δ_H^k and Δ_W^k.

The decoding algorithm can be simply deducted by replacing the TTP encoding algorithm in Algorithm 3 with the corresponding decoding algorithm.

Section 3.3.1 describes the prediction of each type of data used in the proposed event representation. Section 3.3.2 provides information about the setting of the triple thresholds used in the proposed method. Section 3.3.3 describes the variation of LLC-ARES algorithm to provide RA to any time window Δ_{RA}. Finally, Section 3.3.4 presents a coding example.

3.3.1. Prediction

To be able to employ each one of the four algorithm variations, TTP_x, TTP_y, TTP_e, and TTP_L, four types of predictions, $\hat{N}_e^k$, $(\hat{x}_r^k, \hat{y}_r^k)$, $\hat{x}_i^k$, $\hat{L}_\ell$, are computed by using the following set of equations:

$$\hat{N}_e^k = \begin{cases} \tau_e & \text{if } k = 0, \\ N_e^1 & \text{if } k = 1, \\ \frac{N_e^1 + N_e^1}{2} & \text{if } k = 2, \\ \frac{N_e^{k-3} + N_e^{k-2} + 2N_e^{k-1}}{4} & \text{if } k \geq 3. \end{cases} \quad (1)$$

$$(\hat{x}_r^k, \hat{y}_r^k) = \begin{cases} (\frac{H}{2}, \frac{W}{2}) & \text{if } k = 0, \\ (x_1^\kappa, y_1^\kappa + \tau_y) & \text{if } k > \kappa > 0, N_e^\kappa > 0. \end{cases} \quad (2)$$

$$\hat{x}_i^k = \begin{cases} x_{i-1}^k & \text{if } i = 1 \text{ or } \epsilon_{y_i^k} = 0, \\ \frac{x_{i-1}^k + x_{i-2}^k}{2} & \text{if } i = 2, \\ med(\{x_{i-j}\}_{j=1:w_1}) & \text{if } i > 2 \text{ and } |\epsilon_{y_i^k}| < \tau_x, \\ med(\{x_{i-j}\}_{j=1:w_2}) & \text{if } i > 2 \text{ and } |\epsilon_{y_i^k}| \geq \tau_x. \end{cases} \quad (3)$$

$$\hat{L}_\ell = \begin{cases} 2^{7+\lceil log_2 \Delta_{RA} \rceil} & \text{if } \ell = 1, \\ L_{\ell-1} & \text{otherwise.} \end{cases} \quad (4)$$

In (2), the prediction for the spatial information of the first event, e_1^0, in the same-timestamp subsequence S^k, is set as the sensor's centre $(\frac{H}{2}, \frac{W}{2})$, whereas the rest of the values depend on the first event e_1^κ of the previously nonempty same-timestamp subsequence S^κ. In (3), if $\epsilon_{y_i^k}$ is small, $\hat{x}_i^k$ is set as the median of a small prediction window of size w_1; otherwise it is of a larger prediction window of size w_2. In our work, we set the parameters as follows: $\tau_e = 10, \tau_x = 2^3 + 2^4, \tau_y = 3, w_1 = 5, w_2 = 15$.

3.3.2. Threshold Setting

In this paper, the triple threshold parameters, $\Delta_e, \Delta^{e1}, \Delta^S, \Delta_H^{k+1}, \Delta_W^{k+1}$, and Δ^L are selected as power-2 numbers, and are set as follows:

$$\Delta_e = (2^2, 2^2), \quad (5)$$

$$\Delta^{e1} = (2^3, 2^4, 2^5), \quad (6)$$

$$\Delta_H^{k+1} = \begin{cases} \Delta^{e1} & \text{if } k = 0 \\ (2^5, 2^5, 2^6) & \text{if } k > 0 \ \& \ \epsilon^k < 8, \\ (2^4, 2^4, 2^5) & \text{otherwise} \end{cases} \quad (7)$$

$$\Delta_W^{k+1} = \begin{cases} (2^2, 2^3, 2^4) & \text{if } k = 0 \\ (2^1, 2^1, 2^2) & \text{if } k > 0 \ \& \ \epsilon^k < 4 \\ (2^1, 2^2, 2^3) & \text{if } k > 0 \ \& \ \epsilon^k < 8 \\ (2^2, 2^2, 2^3) & \text{if } k > 0 \ \& \ \epsilon^k < 16 \\ (2^2, 2^3, 2^4) & \text{otherwise} \end{cases} \quad (8)$$

$$\Delta^S = (2^8, 2^{10}, 2^{12}), \quad (9)$$

$$\Delta^L = (2^{5+\lceil log_2 \Delta_{RA} \rceil}, 2^{7+\lceil log_2 \Delta_{RA} \rceil}, 2^{9+\lceil log_2 \Delta_{RA} \rceil}). \quad (10)$$

3.3.3. Random Access Functionality

LLC-ARES-RA is an LLC-ARES version which provides RA to any time window of size Δ_{RA}. Hence, $\mathcal{S}_\mathcal{T}$ is now divided into $\mathcal{P} = \lceil \frac{\mathcal{T}}{\Delta_{RA}} \rceil$ packages of Δ_{RA} time-length, denoted $\mathcal{S}_\mathcal{T} = \{\mathcal{S}_\ell\}_{\ell=1,2,\ldots,\mathcal{P}}$. The proposed LLC-ARES is employed to encode each package $\mathcal{S}_\ell$ as the bitstream set $\{\mathcal{B}_{\ell k}\}_{k=0,1,\ldots,\Delta_{RA}-1}$, which is collected as the package ℓ bitstream, $\mathcal{B}_\ell = [\mathcal{B}_{\ell 0} \ \mathcal{B}_{\ell 1} \cdots \mathcal{B}_{\ell \Delta_{RA}}]$, having L_ℓ bit length. The TTP_L version is employed to encode L_ℓ using the prediction $\hat{L}_\ell$, computed using (4), and the two triple threshold Δ^S and Δ^L, and to generate the header bitstream, $\mathcal{B}_\ell^H$, as depicted in Figure 1. Hence, the bitstreams of the set $\{L_\ell\}_{\ell=1,2,\ldots,\mathcal{P}}$ are collected by the header bitstream, denoted as $\mathcal{B}^H = [\mathcal{B}_1^H \ \mathcal{B}_2^H \cdots \mathcal{B}_\mathcal{P}^H]$, whereas all package bitstreams are collected by the sequence bitstream, denoted as $\mathcal{B}^S = [\mathcal{B}_1 \ \mathcal{B}_2 \cdots \mathcal{B}_\mathcal{P}]$. Finally, the compressed file with RA collects the $\mathcal{B}^H$ and $\mathcal{B}^S$ bitstreams in this order.

3.3.4. A Coding Example

Figure 5 presents in detail the workflow of encoding by using the proposed LLC-ARES method an asynchronous event sequence of 2 µs time-length, containing 23 triggered events. The input sequence received from the event sensor is initially represented by using the EE order. The proposed sequence representation is employed by first grouping and then rearranging the asynchronous event sequence by using the ST order. Because the input sequence contains two timestamps, the ST order consist of the same-timestamp subsequence S^0 of 10 events and the same-timestamp subsequence S^1 or 13 events. LLC-ARES encodes each data structure by using different TTP variations as described in Algorithm 3.

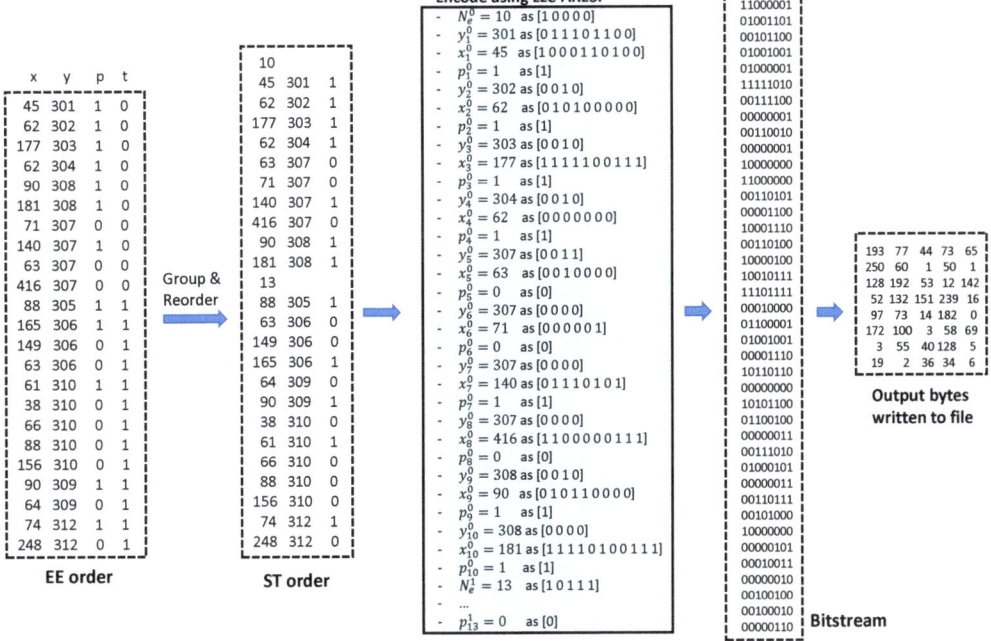

Figure 5. The encoding workflow using the proposed LLC-ARES method as an asynchronous event sequence of 2 µs time-length, containing 23 events. The input sequence, represented by using the EE order, is first grouped and rearranged by using the proposed ST order. LLC-ARES encodes each data structure by using different TTP variations as an output bitstream of 316 bits stored by using 40 bytes, i.e., 40 numbers having an 8-bit representation.

4. Experimental Evaluation

4.1. Experimental Setup

In our work, the experimental evaluation is carried out on large-scale outdoor stereo event camera datasets [33], called DSEC. They contain 82 asynchronous event sequences captured for network training (training data) by using the Prophesee Gen3.1 event sensor placed on top of a moving car, having a $W \times H = 640 \times 480$ pixel resolution. All results reported in this paper use the DSEC asynchronous event sequences sorted in the ascending order of their event acquisition density. By driving at different speeds and in different outdoor scenarios, the DSEC sequences provide a highly variable density of events (see Figure 5a, in which one can see that the event density variates between 5 and 30 Mevps). Figure 6b depicts the cumulated number of events over the first 10 s of the DSEC sequences having the lowest, medium, and highest acquired event density shown in Figure 6a. To limit the runtime of state-of-the-art codecs, for each event sequence, only the first $\mathcal{T} = 10^8$ µs

(100 s) of captured event data are encoded in this work. The DSEC dataset is made publicly available online [34].

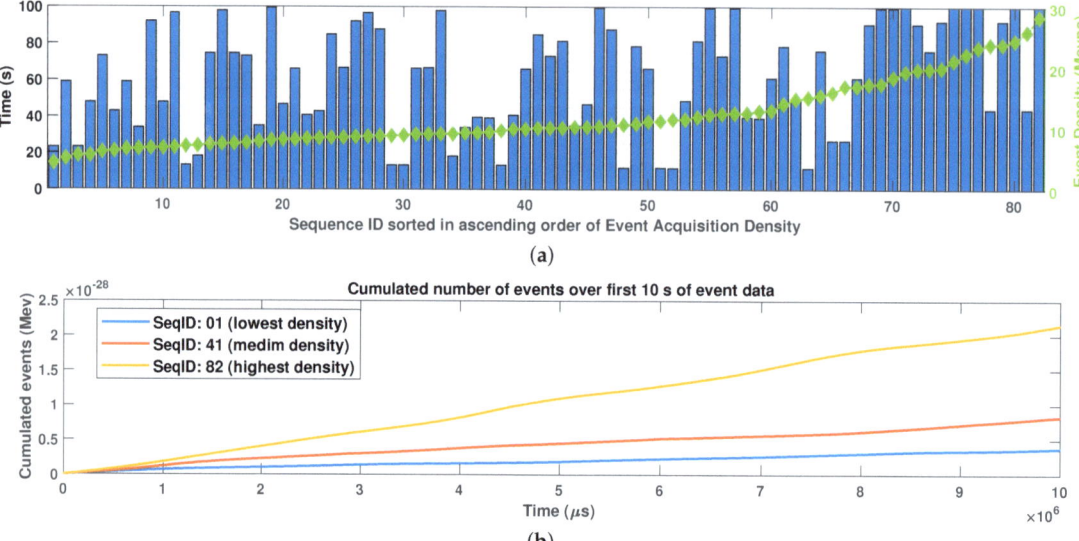

Figure 6. (a) The DSEC sequence time length (s) and event density (Mevps), where the asynchronous event sequences are sorted in ascending order of the sequence acquisition density and the sequence time length was constrained to contain only the first $\mathcal{T} = 10^8$ µs (100 s) of the captured event data. (b) The cumulated number of events (Mev) over the first 10 s of the DSEC sequences having the lowest (SeqID: 01), medium (SeqID: 41), and highest (SeqID: 82) acquired event density.

The proposed method, LLC-ARES, is implemented in the C programming language. The LLC-ARES-RA version is tested by using a time window of Δ_{RA} of 10^2 µs, 10^3 µs, and 10^4 µs, where for each event sequence only the first $\mathcal{T} = 10^7$ µs of captured event data are encoded. The raw data size is computed by using the sensor specifications of 8 B per event.

The compression results are compared by using the following metrics:

(c1) Compression ratio (CR), defined as the ratio between the raw data size and the compressed file size;

(c2) Relative compression (RC), defined as the ratio between the compressed file size of a target codec and the compressed file size of LLC-ARES; and

(c3) Bit rate (BR), defined as the ratio between the compressed file size in bits and the number of events in the asynchronous event sequence, measured in bits per event (bpev), e.g., raw data has 64 bpev.

The runtime results are compared by using the following metrics:

(t1) Event density (ρ_E), defined as the ratio between the number of events in the asynchronous event sequence and the encoding/acquisition time, measured in millions of events per second (Mevps);

(t2) Time ratio (TR), defined as the ratio between the data acquisition time and the codec encoding time; and

(t3) Runtime, defined as the ratio between the encoding/decoding time (µs) and the number of events.

The LLC-ARES performance is compared with the following state-of-the-art traditional data compression codecs:

(a) ZLIB [21] (version 1.2.3 available online [35]);
(b) LZMA [22]; and
(c) Bzip2 (version 1.0.5 available online [36]).

One can note that the comparison with [12] was not possible, as the codec is not publicly available and the dataset is made available only for academic research purposes.

4.2. Compression Results

Figure 7 shows the CR results and Figure 8 shows the BR results over DSEC [34]. One can note that, for state-of-the-art methods, the proposed ST order provides an improved performance of up to 96% compared with the sensor's EE order. LLC-ARES (designed for low-power chip integration) provides an improved performance compared with all state-of-the-art codecs (designed for SoC integration) over the sequences having a small and medium event density, and a close performance over the sequences having a high event density as more complex coding techniques are employed by the traditional lossless data compression methods.

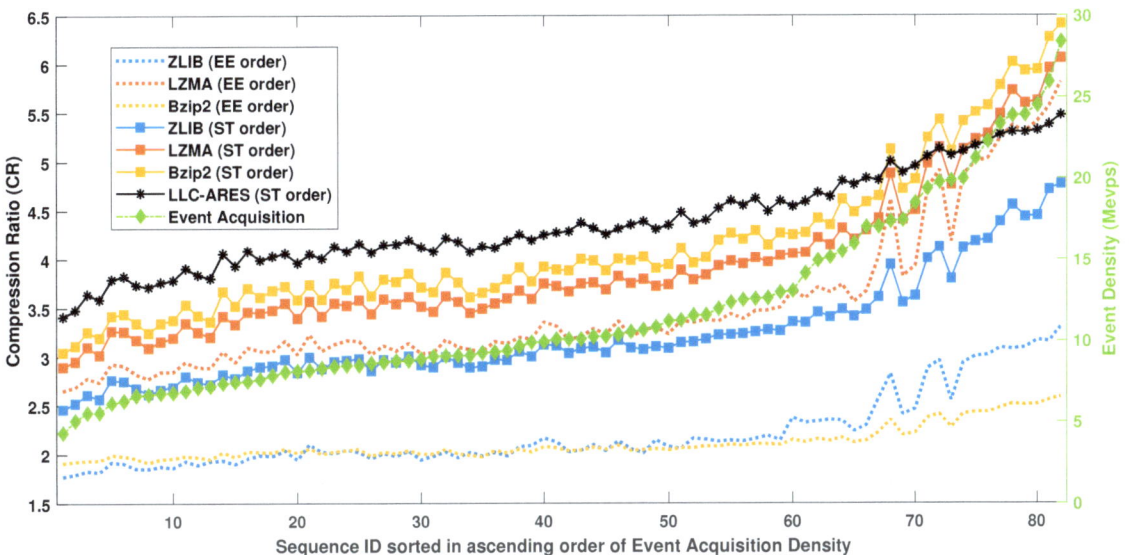

Figure 7. The compression ratio (CR) results over the DSEC dataset [34], where the asynchronous event sequences are sorted in ascending order of the sequence acquisition density.

Table 1 shows the average CR and BR results over DSEC [34]. One can note that, compared with the state-of-the-art performance-oriented lossless data compression codecs, Bzip2, LZMA, and ZLIB, the proposed LLC-ARES codec provides the following:

(i) an average CR improvement of 5.49%, 11.45%, and 35.57%, respectively;
(ii) an average BR improvement of 7.37%, 13.40%, and 37.12%, respectively; and
(iii) an average bitsavings of 1.09 bpev, 1.99 bpev, and 5.50 bpev, respectively.

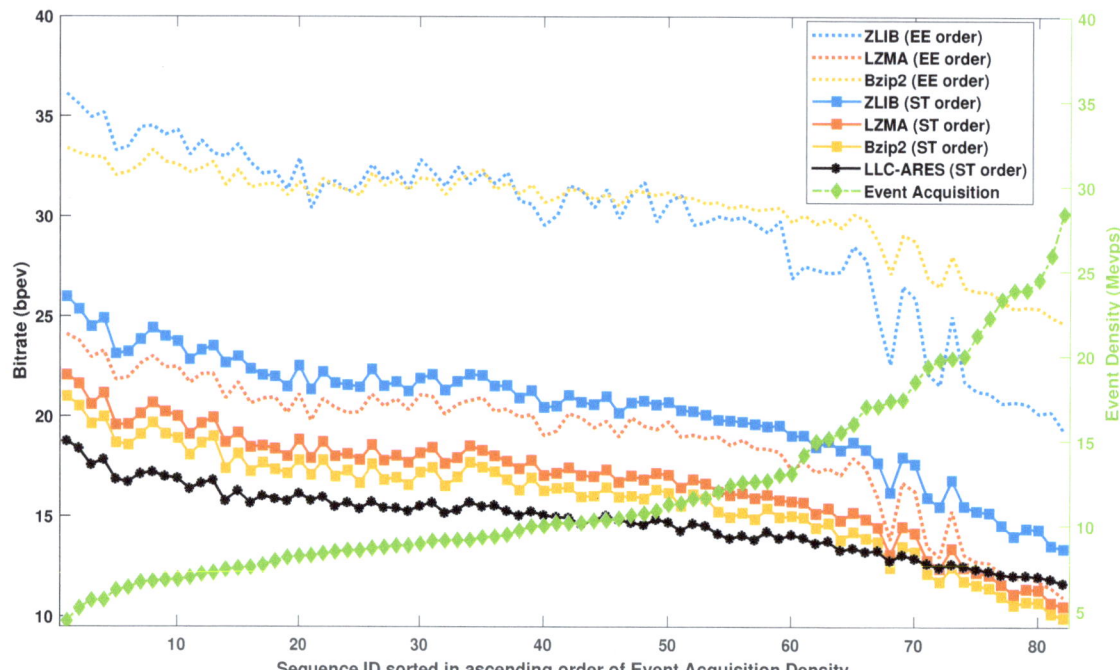

Figure 8. The bitrate (BR) results over DSEC [34], where the asynchronous event sequences are sorted in ascending order of the sequence acquisition density.

Table 1. Average performance over DSEC by using the EE and ST order.

Method		ZLIB [35]	LZMA [22]	bzip2 [36]	Proposed LLC-ARES
CR	EE order	2.21	3.51	2.11	–
	ST order	3.22	3.92	4.14	**4.3**
EBR (bpev)	EE order	29.65	18.91	30.50	–
	ST order	20.32	16.80	15.91	**14.8**
ρ_E (Mevps)	ST order	1.392	0.275	2.453	**5.736**
TR	ST order	0.133	0.027	0.246	**0.531**

4.3. Runtime Results

Figure 9 shows the event density results and Figure 10 shows the TR results over DSEC. One can note that compared with runtime performance of state-of-the-art codecs, LLC-ARES provides a performance much closer to real time for all sequences, and an outstanding performance for the sequences having a high event density. More exactly, LLC-ARES provides a much faster coding speed than the state of the art for the case of high event acquisition density. Whereas the asynchronous event sequences have a very low event acquisition density, LLC-ARES provides an encoding speed as close as approximately 90% of the real-time performance (see Figure 10). Moreover, the software implementation was not optimized, as it can be further improved by a software developer expert to provide an improved runtime performance when deployed on an ESP chip.

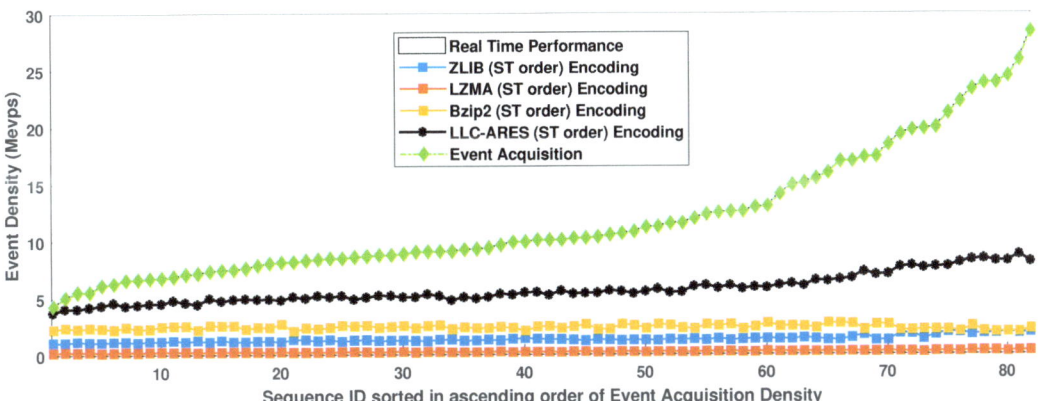

Figure 9. The encoded event density results over the DSEC dataset [34], where the asynchronous event sequences are sorted in ascending order of the sequence acquisition density.

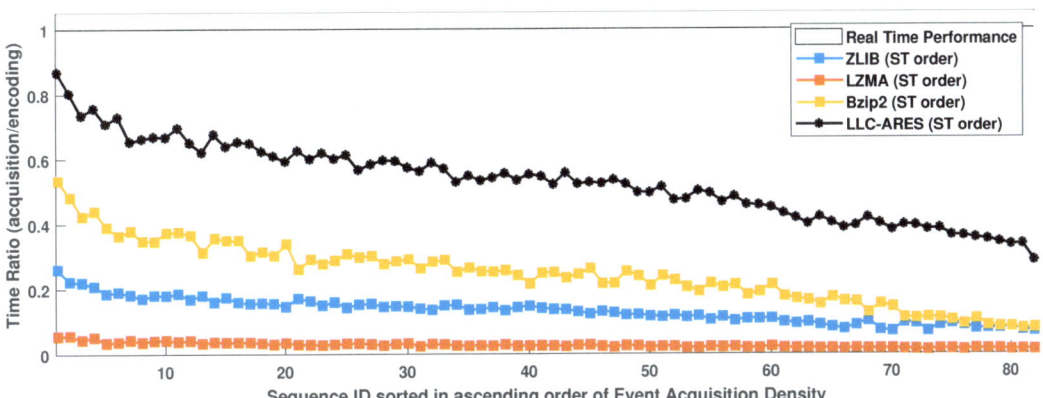

Figure 10. The time ratio (TR) results over the DSEC dataset [34], where the asynchronous event sequences are sorted in ascending order of the sequence acquisition density.

Table 1 shows the average event density and TR results over DSEC. One can note that, compared with the state-of-the-art lossless data compression codecs, Bzip2, LZMA, and ZLIB, the proposed LLC-ARES codec provides the following:

(i) an average event density improvement of $234\times$, $412\times$, and $2086\times$, respectively; and
(ii) an average TR improvement of $216\times$, $401\times$, and $1969\times$, respectively.

Figures 11 and 12 show the encoding and decoding runtime over DSEC, respectively. Note that LLC-ARES is a symmetric codec, wherein the encoder and decoder have similar complexity and runtime, whereas the traditional state-of-the-art lossless data compression methods are asymmetric codecs, as the encoder is much more complex than the decoder. Table 2 presents the average results over DSEC by using the EE order and the proposed ST order. Note that the LLC-ARES performance is approximately 10 μs/ev for both encoding and decoding, while the traditional state-of-the-art lossless data compression methods achieve an encoding time between 135% and 515% higher than LLC-ARES and a decoding time between 92% lower and 58% higher than LLC-ARES.

The implementation of LLC-ARES was not optimized, as the implemented method must be redesigned for integration into low-power chips. These experimental results show that a proof-of-concept implementation of the algorithm on a CPU machine provides an improved performance compared with the state-of-the-art methods when tested on the same experimental setup. Please note that only LLC-ARES employs simple coding techniques so that it is suitable for hardware implementation into low-power ESP chips.

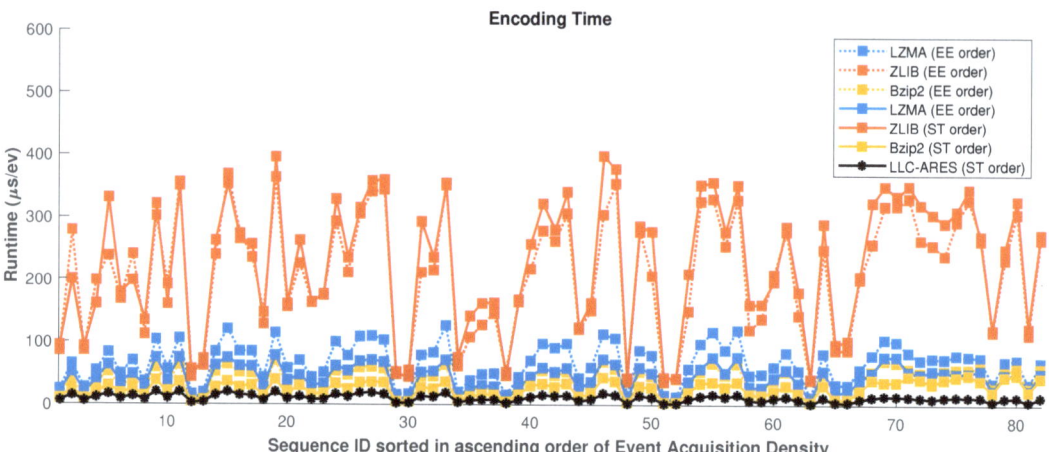

Figure 11. Encoding runtime results over the DSEC dataset [34], where the asynchronous event sequences are sorted in ascending order of the sequence acquisition density.

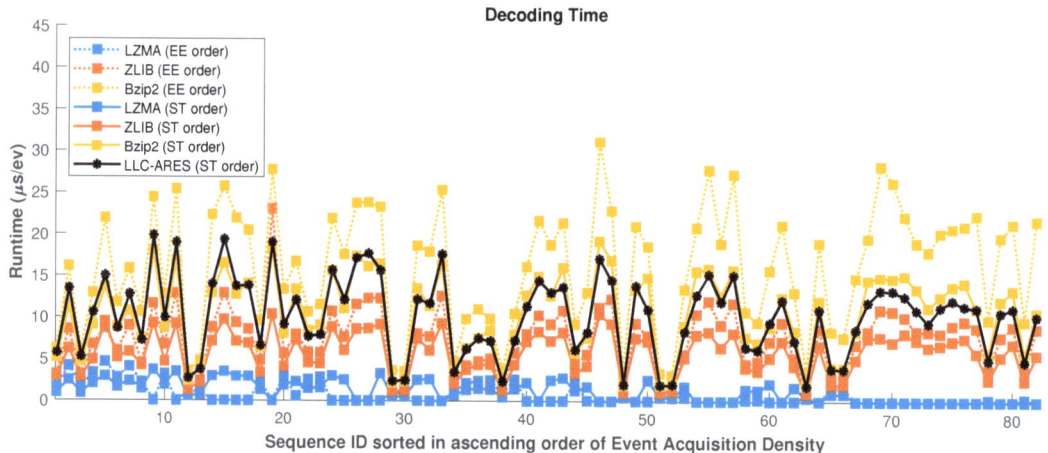

Figure 12. Decoding runtime results over the DSEC dataset [34], where the asynchronous event sequences are sorted in ascending order of the sequence acquisition density.

Table 2. Average runtime results over DSEC using EE and ST order.

Method		ZLIB [35]	LZMA [22]	bzip2 [36]	Proposed LLC-ARES
Encoding Runtime	EE order	67.20 μs/ev	210.39 μs/ev	40.91 μs/ev	–
	ST order	44.70 μs/ev	227.27 μs/ev	25.75 μs/ev	10.92 μs/ev
Decoding Runtime	EE order	**0.78 μs/ev**	7.46 μs/ev	16.09 μs/ev	–
	ST order	1.14 μs/ev	5.71 μs/ev	10.58 μs/ev	10.21 μs/ev

4.4. RA Results

Figure 13 shows the RC results over DSEC. One can note that the RC results are quite similar, as the size of the header bitstream is neglectable compared with the time-window sequence bitstream. When providing RA to the smallest tested time window of

$\Delta_{RA} = 100$ μs, compared with LLC-ARES, the coding performance of the proposed LLC-ARES-RA method decreases with less than 0.19% when the encoded header information is stored in memory and less than 0.35% when the decoded header information is stored in memory, denoted here as memory usage (MU) results.

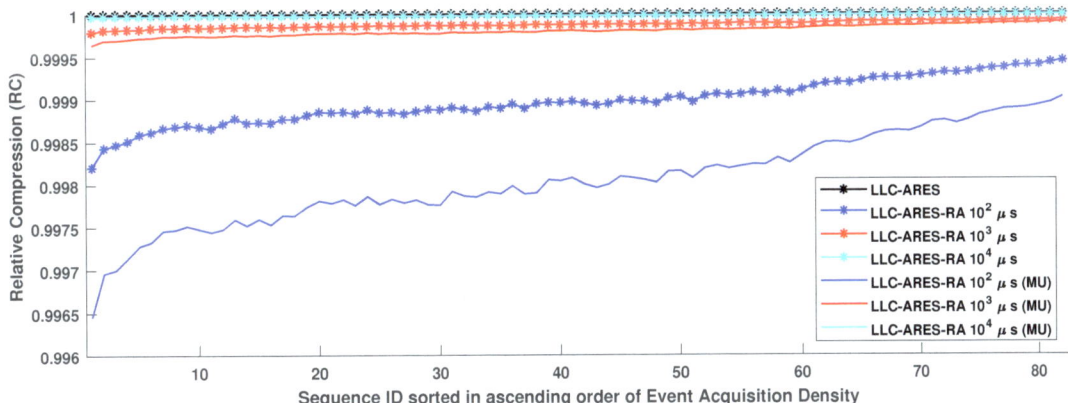

Figure 13. The relative compression (RC) results for RA results over the DSEC dataset [34], wherein the asynchronous event sequences are sorted in ascending order of the sequence acquisition density.

5. Conclusions

In this paper, we proposed a novel lossless compression method for encoding the event data acquired by the new event sensor and represented as an asynchronous event sequence. The proposed LLC-ARES method is built based on a novel low-complexity coding technique so that it is suitable for hardware implementation into low-power ESP chips. The proposed low-complexity coding scheme, TTP, creates short-depth decision trees to reduce either the binary representation of the residual error computed based on a simple prediction, or the binary representation of the true value. The proposed event representation employs the novel ST order, whereby same-timestamp events are first grouped into same-timestamp subsequences, and then reordered to improve the coding performance. The proposed LLC-ARES-RA method provides RA to any time window by employing a header structure to store the length of the bitstream packages.

The experimental results demonstrate that the proposed LLC-ARES codec provides an improved coding performance and a closer to real-time runtime performance compared with state-of-the-art lossless data compression codecs. More exactly, compared with Bzip2 [36], LZMA [22], and ZLIB [35], respectively, the proposed method provides:

(1) an average CR improvement of 5.49%, 11.45%, and 35.57%;
(2) an average BR improvement of 7.37%, 13.40%, and 37.12%;
(3) an average bitsavings of 1.09 bpev, 1.99 bpev, and 5.50 bpev;
(4) an average event density improvement of 234×, 412×, and 2086×; and
(5) an average TR improvement of 216×, 401×, and 1969×.

To our knowledge, the paper proposes the first low-complexity lossless compression method for encoding asynchronous event sequences that is suitable for hardware implementation into low-power chips.

Author Contributions: Conceptualization, I.S.; methodology, I.S.; software, I.S.; validation, I.S.; formal analysis, I.S.; investigation, I.S.; writing—original draft preparation, I.S.; writing—review and editing, I.S. and R.C.B.; visualization, I.S.; supervision, R.C.B. All authors have read and agreed to the published version of the manuscript.

Funding: This research received no external funding.

Institutional Review Board Statement: Not applicable.

Informed Consent Statement: Not applicable.

Data Availability Statement: Not applicable.

Conflicts of Interest: The authors declare no conflict of interest.

Abbreviations

The following abbreviations are used in this manuscript:

DVS	Dynamic Vision Sensor
APS	Active Pixel Sensor
DAVIS	Dynamic and Active-pixel VIsion Sensor
EF	Event Frame
RA	Random Access
TALVEN	Time Aggregation-based Lossless Video Encoding for Neuromorphic sensor
ESP	Event Signal Processing
SoC	System-on-a-chip
EMI	Event Map Image
CPV	Concatenated Polarity Vector
HEVC	High Efficiency Video Coding
SNN	Spike Neural Network
EGC	Elias-Gamma-Coding
LLC-ARES	Low-Complexity Lossless AsynchRonous Event Sequences
LLC-ARES-RA	LLC-ARES with RA
ZLIB	Zeta Library
LZMA	Lempel–Ziv–Markov chain Algorithm
G-PCC	Geometry-based Point Cloud Compression
CR	Compression Ratio
BR	Bitrate
TR	Time Ratio

References

1. Lichtsteiner, P.; Posch, C.; Delbruck, T. A 128× 128 120 dB 15 µs Latency Asynchronous Temporal Contrast Vision Sensor. *IEEE J. Solid State Circ.* **2008**, *43*, 566–576. [CrossRef]
2. Brandli, C.; Berner, R.; Yang, M.; Liu, S.C.; Delbruck, T. A 240 × 180 130 dB 3 µs Latency Global Shutter Spatiotemporal Vision Sensor. *IEEE J. Solid State Circ.* **2014**, *49*, 2333–2341. [CrossRef]
3. Pan, L.; Scheerlinck, C.; Yu, X.; Hartley, R.; Liu, M.; Dai, Y. Bringing a Blurry Frame Alive at High Frame-Rate With an Event Camera. In Proceedings of the 2019 IEEE/CVF Conference on Computer Vision and Pattern Recognition (CVPR), Long Beach, CA, USA, 15–20 June 2019; pp. 6813–6822. [CrossRef]
4. Gehrig, D.; Rebecq, H.; Gallego, G.; Scaramuzza, D. EKLT: Asynchronous Photometric Feature Tracking using Events and Frames. *Int. J. Comput. Vis.* **2020**, *128*, 601–618. [CrossRef]
5. Iaboni, C.; Lobo, D.; Choi, J.W.; Abichandani, P. Event-Based Motion Capture System for Online Multi-Quadrotor Localization and Tracking. *Sensors* **2022**, *22*, 3240. [CrossRef] [PubMed]
6. Zhu, A.; Yuan, L.; Chaney, K.; Daniilidis, K. EV-FlowNet: Self-Supervised Optical Flow Estimation for Event-based Cameras. In Proceedings of the Robotics: Science and Systems, Pittsburgh, PA, USA, 26–30 June 2018. [CrossRef]
7. Brandli, C.; Mantel, T.; Hutter, M.; Höpflinger, M.; Berner, R.; Siegwart, R.; Delbruck, T. Adaptive pulsed laser line extraction for terrain reconstruction using a dynamic vision sensor. *Front. Neurosci.* **2014**, *7*, 1–9. [CrossRef]
8. Li, S.; Feng, Y.; Li, Y.; Jiang, Y.; Zou, C.; Gao, Y. Event Stream Super-Resolution via Spatiotemporal Constraint Learning. In Proceedings of the 2021 IEEE/CVF International Conference on Computer Vision (ICCV), Montreal, QC, Canada, 11 October 2021; pp. 4460–4469. [CrossRef]

9. Yu, Z.; Zhang, Y.; Liu, D.; Zou, D.; Chen, X.; Liu, Y.; Ren, J. Training Weakly Supervised Video Frame Interpolation with Events. In Proceedings of the 2021 IEEE/CVF International Conference on Computer Vision (ICCV), Montreal, QC, Canada, 10–17 October 2021; pp. 14569–14578. [CrossRef]
10. Wang, Y.; Yang, J.; Peng, X.; Wu, P.; Gao, L.; Huang, K.; Chen, J.; Kneip, L. Visual Odometry with an Event Camera Using Continuous Ray Warping and Volumetric Contrast Maximization. *Sensors* **2022**, *22*, 5687. [CrossRef]
11. Gallego, G.; Delbrück, T.; Orchard, G.; Bartolozzi, C.; Taba, B.; Censi, A; Leutenegger, S.; Davison, A.; Conradt, J.; Daniilidis, K.; et al. Event-Based Vision: A Survey. *IEEE Trans. Pattern Anal. Mach. Intell.* **2022**, *44*, 154–180. [CrossRef]
12. Bi, Z.; Dong, S.; Tian, Y.; Huang, T. Spike Coding for Dynamic Vision Sensors. In Proceedings of the Data Compression Conf., Snowbird, UT, USA, 27–30 March 2018; pp. 117–126. [CrossRef]
13. Dong, S.; Bi, Z.; Tian, Y.; Huang, T. Spike Coding for Dynamic Vision Sensor in Intelligent Driving. *IEEE Internet Things J.* **2019**, *6*, 60–71. [CrossRef]
14. Khan, N.; Iqbal, K.; Martini, M.G. Lossless Compression of Data From Static and Mobile Dynamic Vision Sensors-Performance and Trade-Offs. *IEEE Access* **2020**, *8*, 103149–103163. [CrossRef]
15. Khan, N.; Iqbal, K.; Martini, M.G. Time-Aggregation-Based Lossless Video Encoding for Neuromorphic Vision Sensor Data. *IEEE Internet Things J.* **2021**, *8*, 596–609. [CrossRef]
16. Banerjee, S.; Wang, Z.W.; Chopp, H.H.; Cossairt, O.; Katsaggelos, A.K. Lossy Event Compression Based On Image-Derived Quad Trees And Poisson Disk Sampling. In Proceedings of the IEEE Int. Conf. Image Process., Imaging Without Borders, Anchorage, AK, USA, 19–22 September 2021; pp. 2154–2158. [CrossRef]
17. Schiopu, I.; Bilcu, R.C. Lossless Compression of Event Camera Frames. *IEEE Signal Process. Lett.* **2022**, *29*, 1779–1783. [CrossRef]
18. Schiopu, I.; Bilcu, R.C. Low-Complexity Lossless Coding for Memory-Efficient Representation of Event Camera Frames. *IEEE Sens. Lett.* **2022**, *6*, 1–4. [CrossRef]
19. Elias, P. Universal codeword sets and representations of the integers. *IEEE Trans. Inf. Theory* **1975**, *21*, 194–203. [CrossRef]
20. Ziv, J.; Lempel, A. A universal algorithm for sequential data compression. *IEEE Trans. Inf. Theory* **1977**, *23*, 337–343. [CrossRef]
21. Deutsch, P.; Gailly, J.L. Zlib Compressed Data Format Specification; Version 3.3; RFC: 1950; IETF. 1996. Available online: https://www.ietf.org/ (accessed on 19 July 2021).
22. Pavlov, I. LZMA SDK (Software Development Kit). Available online: https://www.7-zip.org/ (accessed on 19 July 2021).
23. Burrows, M.; Wheeler, D.J. *A Block-Sorting Lossless Data Compression Algorithm*; IEEE: Piscataway, NJ, USA, 1994.
24. Martini, M.G.; Adhuran, J.; Khan, N. Lossless Compression of Neuromorphic Vision Sensor Data based on Point Cloud Representation. *IEEE Access* **2022**, *10* 121352–121364. [CrossRef]
25. Henri Rebecq, T.H.; Scaramuzza, D. Real-time Visual-Inertial Odometry for Event Cameras using Keyframe-based Nonlinear Optimization. In Proceedings of the British Machine Vision Conference (BMVC), London, UK, 21–24 November 2017; Tae-Kyun, K., Stefanos Zafeiriou, G.B., Mikolajczyk, K., Eds.; BMVA Press: Durham, UK; pp. 16.1–16.12. [CrossRef]
26. Maqueda, A.I.; Loquercio, A.; Gallego, G.; Garcia, N.; Scaramuzza, D. Event-Based Vision Meets Deep Learning on Steering Prediction for Self-Driving Cars. In Proceedings of the 2018 IEEE/CVF Conference on Computer Vision and Pattern Recognition, Salt Lake City, UT, USA, 18–23 June 2018; IEEE: Piscataway, NJ, USA. [CrossRef]
27. Almatrafi, M.; Baldwin, R.; Aizawa, K.; Hirakawa, K. Distance Surface for Event-Based Optical Flow. *IEEE Trans. Pattern Anal. Mach. Intell.* **2020**, *42*, 1547–1556. [CrossRef]
28. Benosman, R.; Clercq, C.; Lagorce, X.; Ieng, S.H.; Bartolozzi, C. Event-Based Visual Flow. *IEEE Trans. Neural Netw. Learn. Syst.* **2014**, *25*, 407–417. [CrossRef]
29. Zhu, A.; Yuan, L.; Chaney, K.; Daniilidis, K. Unsupervised Event-Based Learning of Optical Flow, Depth, and Egomotion. In Proceedings of the 2019 IEEE/CVF Conference on Computer Vision and Pattern Recognition (CVPR), Los Alamitos, CA, USA, 15–20 June 2019; IEEE Computer Society: Washington, DC, USA, 2019; pp. 989–997. [CrossRef]
30. Baldwin, R.; Liu, R.; Almatrafi, M.M.; Asari, V.K.; Hirakawa, K. Time-Ordered Recent Event (TORE) Volumes for Event Cameras. *IEEE Trans. Pattern Anal. Mach. Intell.* **2022**, *Early Access*. [CrossRef]
31. Sullivan, G.J.; Ohm, J.R.; Han, W.J.; Wiegand, T. Overview of the High Efficiency Video Coding (HEVC) Standard. *IEEE Trans. Circ. Syst. Video Technol.* **2012**, *22*, 1649–1668. [CrossRef]
32. Zhu, L.; Dong, S.; Huang, T.; Tian, Y. Hybrid Coding of Spatiotemporal Spike Data for a Bio-Inspired Camera. *IEEE Trans. Circ. Syst. Video Technol.* **2021**, *31*, 2837–2851. [CrossRef]
33. Gehrig, M.; Aarents, W.; Gehrig, D.; Scaramuzza, D. DSEC: A Stereo Event Camera Dataset for Driving Scenarios. *IEEE Robot. Autom. Lett.* **2021**, *6*, 4947–4954. [CrossRef]
34. DSEC Dataset. Available online: https://dsec.ifi.uzh.ch/dsec-datasets/download/ (accessed on 1 October 2021).
35. Vollan, G. ZLIB Pre-Build DLL. Available online: http://www.winimage.com/zLibDll/ (accessed on 19 July 2021).
36. Seward, J. bzip2 Pre-Build Binaries. Available online: http://gnuwin32.sourceforge.net/packages/bzip2.htm (accessed on 19 July 2021).

Article

Learning-Based Rate Control for High Efficiency Video Coding

Sovann Chen [1], Supavadee Aramvith [2,*] and Yoshikazu Miyanaga [3]

1. Department of Electrical Engineering, Faculty of Engineering, Chulalongkorn University, Bangkok 10330, Thailand
2. Multimedia Data Analytics and Processing Research Unit, Department of Electrical Engineering, Faculty of Engineering, Chulalongkorn University, Bangkok 10330, Thailand
3. Chitose Institute of Science and Technology, Chitose 066-8655, Japan
* Correspondence: supavadee.a@chula.ac.th

Abstract: High efficiency video coding (HEVC) has dramatically enhanced coding efficiency compared to the previous video coding standard, H.264/AVC. However, the existing rate control updates its parameters according to a fixed initialization, which can cause errors in the prediction of bit allocation to each coding tree unit (CTU) in frames. This paper proposes a learning-based mapping method between rate control parameters and video contents to achieve an accurate target bit rate and good video quality. The proposed framework contains two main structural codings, including spatial and temporal coding. We initiate an effective learning-based particle swarm optimization for spatial and temporal coding to determine the optimal parameters at the CTU level. For temporal coding at the picture level, we introduce semantic residual information into the parameter updating process to regulate the bit correctly on the actual picture. Experimental results indicate that the proposed algorithm is effective for HEVC and outperforms the state-of-the-art rate control in the HEVC reference software (HM-16.10) by 0.19 dB on average and up to 0.41 dB for low-delay P coding structure.

Keywords: HEVC; learning-based rate control; PSO

Citation: Chen, S.; Aramvith, S.; Miyanaga, Y. Learning-Based Rate Control for High Efficiency Video Coding. *Sensors* **2023**, *23*, 3607. https://doi.org/10.3390/s23073607

Academic Editors: Ittetsu Taniguchi, Jinjia Zhou and Xin Jin

Received: 17 February 2023
Revised: 24 March 2023
Accepted: 27 March 2023
Published: 30 March 2023

Copyright: © 2023 by the authors. Licensee MDPI, Basel, Switzerland. This article is an open access article distributed under the terms and conditions of the Creative Commons Attribution (CC BY) license (https://creativecommons.org/licenses/by/4.0/).

1. Introduction

Multimedia technology has been upgraded from one generation to another to fulfill daily needs such as television, telephones, computers, robots, etc. Numerous multimedia applications have been utilized, including digital versatile disc (DVD), digital television (TV) broadcasting, video telephony, video teleconferencing, video games, and other forms of video-on-demand. According to [1], the resolution of television broadcasting has been upgraded from standard-definition television (SDTV) to 8K ultra high definition (UHD), which requires a very high bit rate to transmit or store. Furthermore, the video demand on internet traffic is increasing, based on a statistical report in the "Cisco Annual Internet Report (2018–2023)", a Cisco White Paper in 2018 in [2]. Thus, it strongly needs an effective video coding technique to reduce the network traffic load with good visual quality and a lower bit rate.

In general, video properties have four redundancy criteria: spatial redundancy, temporal redundancy, perceptual redundancy, and statistical redundancy, which can be eliminated by the video coding standard [3]. High efficiency video coding (HEVC) [4], an advanced video coding standard released in 2013 by ITU-T and ISO/IEC, can effectively remove the digital video redundancies and achieve a bit rate saving of about fifty percent at the same visual quality by comparing with the previous standard (H.264/AVC [3,5,6]). HEVC is built following the structure of the successful block-based hybrid video coding approach [7], the same as the H.264/AVC video coding standard. In addition, several advanced techniques are applied in HEVC to get efficient compressions, such as flexible partitioning using quadtree structure, prediction modes [8], sample adaptive offset (SAO) [9], and the cutting-edge interpolation technique [10].

Moreover, HEVC needs to have a functional encoder control, known as rate control, to determine the optimum codec parameters to accomplish minimal rate–distortion (R–D) score [11]. Many codec parameters include modes selection, quad-tree structure, motion estimation, and quantization parameter (QP). In common, the rate control algorithms [11,12] are used to define the bit allocation and QP by fixing the other parameters to accomplish the target bit with consistent visual quality. Specifically, rate control needs to manipulate the number of bits from a constant bit rate (CBR) into each coding level, including the group of picture (GOP) level, picture level, and basic units known as macroblocks (MBs) in H.264/AVC. The QP is then regulated to achieve the pre-allocated bits for each coding level, where the larger number of QP leads to a smaller number of allocated bits and vice versa. Encoder controls typically implement a uniform bit allocation in a GOP structure and initialize the fixed encoding parameters for any video contents to preserve a short-term constant output bit rate in the CBR channel. As a result, this implementation faces an infeasible problem of accurately adjusting encoding parameters for each GOP frame. Accordingly, if the target bit is less than the output bits, the encoded bits will rack up in the encoder buffer, causing a buffer overflow. The target bit is greater than the output bits, which implies the buffer underflow. Hence, controlling the relationship between bit rate and QP is essential for maintaining picture quality throughout the video sequence, as buffer overflows and underflows have an undesired effect on video quality fluctuations. Q-domain rate control is a direct estimation that attempts to model a correlation function between bit rate and quantization; the bit allocation can be computed from the QP to allocate for residual information but not for non-residual information. This model can work well when the coding parameters are not very flexible. Another rate control algorithm called ρ-domain rate control is developed [12,13] by introducing a linear function that outputs the coding bit rate from the percentage of zeros among the quantized transform coefficient. The model is effective only if the size of the transform is fixed. Both Q-domain and ρ-domain rate controls are designed to assume a high correlation between bit rate and quantization. This assumption is not valid for the current video codec because the codec becomes progressively variable [4]. Thus, a robust rate control [11], named R–λ rate control, has been released to achieve the best balance between bit rate and distortion. This rate control attempts to improve the coding efficiency and rate control accuracy by using the Lagrangian method, λ, for rate–distortion optimization (RDO).

Although the aim of R–λ rate control is for HEVC to enhance the coding efficiency compared with the conventional methods, two difficulties still need to be solved in HEVC reference software [14], including inaccurate bit allocation and inaccurate λ estimation. For the bit allocation part, the bit consumption of each CU of the first picture is computed by applying one to all initial encoder parameters at the basic unit level. In other words, all CUs are encoded using the same rate control parameters as the picture level. In such a case, the rate control will cause a bit consumption imbalance in the CU due to the spatial characteristic of each CU and result in the error bits' distribution affecting the overall quality control. In addition, the inaccurate bit consumption at each coding level affects the λ adjustment to accomplish the frame bit budget because λ and the bit allocation are highly correlated. Specifically, according to the previous encoding results and the statistical characteristics of the input source data, the encoder parameters are empirically inaccurate, resulting from performance degradation at scene changes.

Based on these considerations, we propose a learning-based mapping method between R–λ parameters and video content to achieve accurate target bit rates and preserve good video quality. We use a feedback re-encoding method for the intra-picture and inter-picture to distribute R–λ parameters adaptively related to picture pattern changes. Additionally, the convolutional neural network (CNN) model [15] is used to capture the powerful spatial representation of the local coding tree units (CTUs). This CNN model is trained on the ImageNet dataset [16]. By incorporating the CNN model with the R–λ rate control algorithm, we can accurately obtain the expected number of bits per CTU. Our problem is a constrained optimization problem, where, by obtaining the optimal encoder control parameters to minimize the distortion subject to a constraint, the actual bit rate consumption is less than the target bit rate. To solve the constrained optimization problem, there are two optimization methods, namely the gradient-based method [17,18] and the non-gradient-based method (known as the evolutionary algorithm) [19–24]. The gradient-based method is effective only when the constraints and objective or penalty function can be derived. Since our model aims to map the high-dimensional feature space of the CTU to the R–λ parameter with the goal of R–D optimization, which cannot directly derive the gradient information from the penalty function, the evolutionary algorithm (EA) is chosen to optimize the parameters of our model. There are several EAs such as evolution strategies (ES) [19], simulated annealing (SA) [20], genetic algorithm (GA) [21], and particle swarm optimization (PSO) [22]. Due to the simplicity and convergence speed characteristics of all EAs [24], PSO is the most powerful one and has been successfully implemented to solve various constrained optimization problems [25–28]. Comprehensively, PSO takes the value of the objective function and uses primitive mathematical operators to solve the social behavior of model parameters. Therefore, PSO is implemented in our model to find the best solution for mapping the characteristics of CTU and rate control parameters. Furthermore, we feed the semantic residue information to adjust the current parameters of rate control updating cross-picture. The main contributions of this paper can be summarized in three aspects:

(i) We propose a learning-based neural network to define the mapping between video contents and rate control parameters to assign CTU budgets correctly;
(ii) We introduce a particle swarm optimization algorithm to finalize the optimal parameters at the basic unit level to maintain the bit budget and obtain good visual video quality;
(iii) We enhance the rate control parameter updating by considering the semantic residue information of the actual inter-picture into rate control.

The rest of the paper is organized as follows. In the next section, we briefly summarize related work. Then, the learning-based parameters of R–λ are described. After that, the experimental results are given. Finally, concluding remarks are provided.

2. Related Works

In this section, we briefly review the existing rate control models: R–Q model, ρ-domain-based Rate-GOP, R–λ models, and deep learning based rate control.

2.1. R–Q Model

The R–Q model [29] has extended to HEVC encoder control, known as a pixel-wise unified R–Q model (URQ); the quadratic R–Q model is defined as in (1),

$$R = aQ^{-1} + bQ^{-2} \tag{1}$$

where R presents as the target bit rate, Q is the quantization parameter, and a and b are the parameters related to the video characteristic. The bit allocation of the URQ model is proposed similarly to the rate control model in H.264/AVC, where the target bit is computed based on the mean absolute difference (MAD) corresponding to the quantization step. As a result, compared with the earlier HEVC video coding standard (HM6.0) [14], the visual quality of the URQ model is slightly improved. However, some issues have been discussed regarding Q-domain rate control [30,31], such that QP is an integer data type that may not be adjusted accurately to achieve a bit budget.

2.2. ρ-Domain-Based Rate-GOP

The enhanced R–Q model known as ρ-domain-based Rate-GOP is proposed in [32] by presenting a new relationship one-to-one quantized transform coefficient with target bit rate. It is formulated as in (2):

$$R_i = \theta_i(1 - \rho_i) \tag{2}$$

where θ_i and ρ_i denote a parameter related to the video pattern and the percentage of zero transform coefficients of frame i, respectively. Additionally, the mapping between non-zero transform coefficients and QP is determined following the quadratic function to properly allocate the bit to non-zero transform units. Consequently, the ρ-domain-based Rate-GOP can significantly achieve better video quality than the Q-domain rate control. Although this indirect relationship between R and Q technique is advantageous, it is still difficult to adapt its estimation to the variable block size transform in HEVC.

2.3. R–λ Model

To overcome the limitations of the R–Q model mentioned above, a new type of encoder control with the hierarchical bit allocation for every picture in a GOP is proposed in [11], called R–λ rate control. First, the author proposed a hyperbolic function as a model to express the characteristics of the R–D relationship, as in (3):

$$D(R) = C \cdot R^{-K} \tag{3}$$

where C and K are parameters related to video content. Then, to minimize (3), λ is determined as the slope of the model in (4).

$$\lambda = -\frac{\partial D}{\partial R} = C \cdot K \cdot R^{-K-1} \tag{4}$$

$$\Leftrightarrow \lambda = \alpha \cdot R^\beta \equiv \gamma \cdot D^\tau. \tag{5}$$

Therefore, λ can indicate the trade-off between bit rate and distortion. If λ is large, the lower bit rate will cause higher distortion. On the other hand, small λ results in a higher bit rate with lower distortion. In addition, a hierarchical bit allocation method [33] is used to allocate different picture weights corresponding to each picture position in the GOP to improve coding efficiency. Furthermore, the QP can be computed by giving λ for each coding level as in (6).

$$QP = 4.2005 \cdot ln(\lambda) + 13.7122. \tag{6}$$

The rate control can obtain stable buffer occupation and codec improvements through the hierarchical bit allocation method and the novel relationship between λ and R. As a result, R–λ rate control is generally used in the advanced video coding standard. However, the R–λ model mainly considers the bit rate by ignoring the characteristics of the video content. Furthermore, the model initializes its parameters by sharing the same fixed constant from the frame to all CTU levels. These aspects can cause video quality degradation.

A distortion-based Lagrange multiplier is proposed in [34] to enhance the compressed video quality in HEVC. The authors used the equivalent of distortion D and λ instead of R–λ. Two main objective functions control the λ adjustment: mean square error (MSE) and absolute error. MSE is calculated from the original and reconstructed video content, while the absolute error is computed by subtracting between the actual and target bit budget. This technique is designed for the non-hierarchical structure of rate control. It can enhance the video quality by an average of 0.23 dB in the low-delay P configuration compared with non-hierarchical R–λ rate control. The R–λ model with a hierarchical structure achieves a higher video quality of 0.26 dB than the R–λ model without a hierarchical structure [11]. This ability of the hierarchical structure in R–λ makes it a common approach as the default HEVC general test condition in [35]. A video quality enhancement of the compressed video worked on R–λ with a hierarchical structure is proposed in [36]. The authors introduced a simple rate control parameter-sharing in a GOP structure (PS-GOP), achieving a higher video quality of 0.07 dB on average and up to 0.17 dB compared to the default HEVC reference software (HM-16.10) [14].

An inter-block dependency-based CTU-level rate control for HEVC is established in [37], known as the RCA model. This proposed RCA is inspired by the temporal-dependent RDO, which is formulated as the fusion between inter-block dependency and

R–D characteristics. This proposed model has achieved a considerable PNSR enhancement. However, the spatial coding units have not been taken into consideration, which would result in parameter propagation errors at the early stage.

2.4. Deep Learning-Based Rate Control

A deep reinforcement learning-based rate control for the dynamic video sequences is designed in [38] to capture the experience gained from the various factors, including brightness, variance, and gradient of each coding unit during the coding process. The proposed model is structured following the Markov decision process in a continuous discrete space to obtain better PSNR and lower-quality fluctuation. Nevertheless, the reinforcement approach has limitations, including a high number of interactions required to learn an optimal policy and difficulty generalizing to new, unseen environments.

Under a random access configuration, a deep convolution features-driven rate control for the HEVC encoders is proposed [39]. The method involves utilizing a pre-trained VGG-16 model to extract perceptual features, which addresses the problem of the rate control estimation. However, the model has not generalized the visual characteristic mapping to the rate control parameter.

Hence, we propose effective R–λ parameters associated with the video content to improve the compressed video quality and maintain the bit budgets at the encoder side. The following section presents the proposed framework in detail.

3. Learning-Based Rate Control

This section introduces a learning-based rate control algorithm, which creates a regression map for the R–λ parameter. The proposed framework is designed, as shown in Figure 1. The green boxes represent the modification rate control model using the feature translation technique and the convolution feature map. First, the input video is fed into the convolution feature map to extract the high dimensional feature space, which contains essential features representing the CTU in the scene. Then, the proposed model learns to translate the input feature space to rate control parameters to get the optimal trade between the target bit rate and distortion rate. Additionally, the dashed lines from the inter- and intra-prediction are indicated to send the convolution feature representation of the video coding with the coding mode, whether intra- or inter-prediction to the Encoder Control block. Figure 2 shows the convolution feature map module and the regression map representations module, which are constructed to generate the R–λ parameters. The regression map is designed as learning-based particle swarm optimization (LB-PSO). Furthermore, the parameter updating for inter-coding is performed by considering residue information. The details of each part are presented in the following subsections.

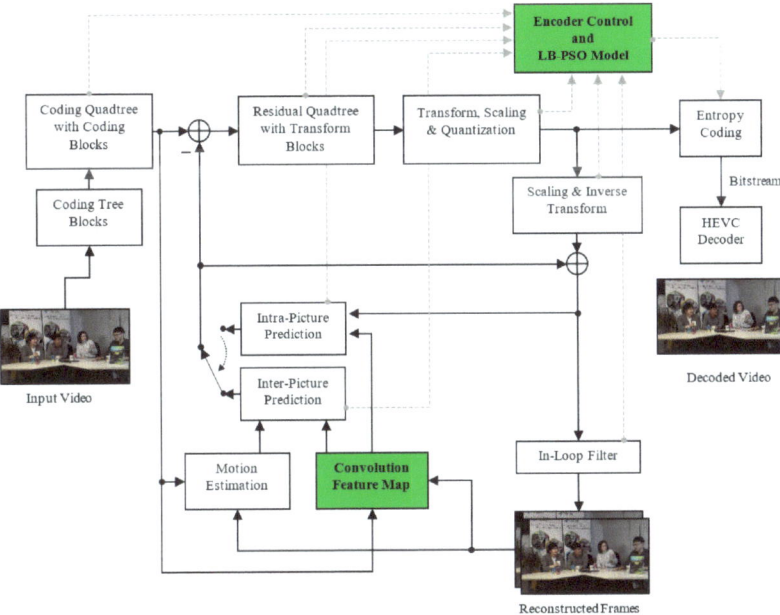

Figure 1. Learning-Based Rate Control Diagram for High Efficiency Video Coding.

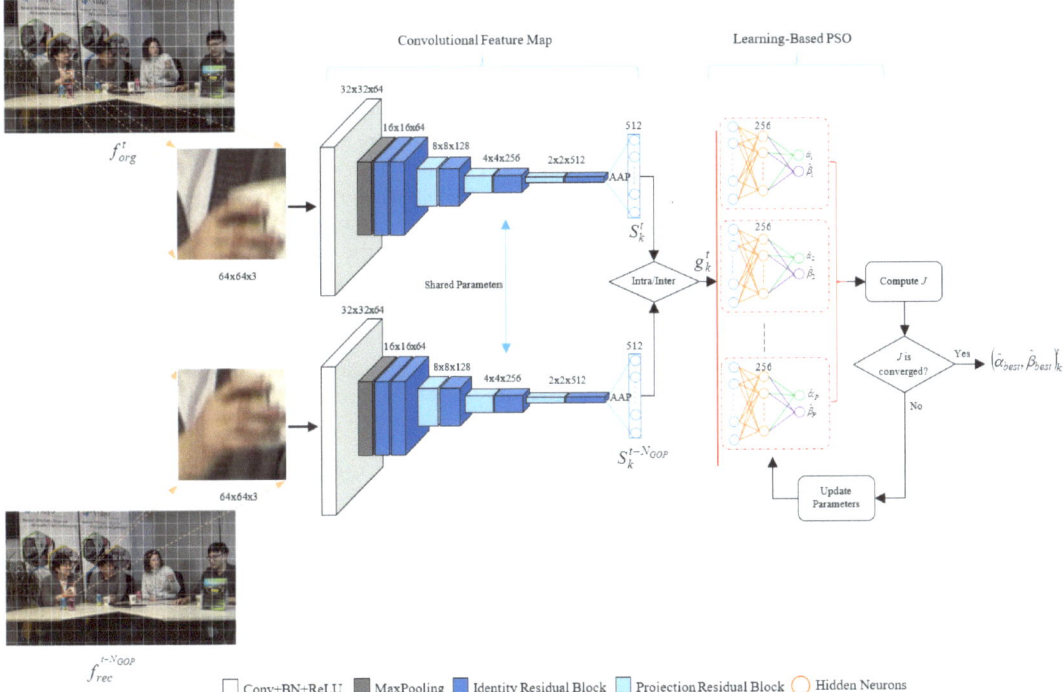

Figure 2. Overview of proposed learning-based particle swarm optimization.

3.1. Convolutional Feature Map

The convolutional feature map (fully convolutional networks—FCNs) is introduced at the first stage to obtain the meaningful spatial representation of CTU pictures for the input of our LB-PSO model. In general, the early layers of convolutions in the deep convolutional networks demonstrate the input image's local or low-level feature information. In contrast, the deeper layers of convolutions indicate the high-level feature information that provides more global information about the image [40]. Additionally, the last fully connected (FC) layer of deep nets is designed to define the high-level feature information into object classes. Since FCNs do not include the FC layer, a relationship between the input image and the final feature output layer is preserved and expressed as data compression, which encodes the raw-pixel representation of the input image to high-level information. This information provides the global feature G representing the input image characteristic. G is fed into our LB-PSO model to generate the R–λ parameters. A pre-trained residual networks (ResNets) [15] model without the FC layer is used to extract the powerful convolutional feature. However, the original input size of ResNets is incompatible with the maximum size of CTUs. The adaptive average pooling (AAP) is then applied to the last convolution layers to ensure the compatibility of input and output dimensions. Figure 2 demonstrates the overall layout of our convolutional feature map architecture.

Suppose a tth frame contains a total K CTUs, then $G^t = \{g_0, g_1, \ldots, g_K\}_t$. Precisely, G is a parameter representing the high-dimensional features required as input to the proposed LB-PSO model. To obtain G for re-feedback coding of each coding structure in HEVC, i.e., intra- or inter-pictures, we define G as in (7):

$$g_k^t = \begin{cases} S_k^t, & \text{if intra-picture.} \\ \left| S_k^t - S_k^{t-N_{GOP}} \right|_{(t \bmod c)}, & \text{otherwise.} \end{cases} \quad (7)$$

where $k \in K$, and c ($c > 0$) is a constant to determine the frame index for re-feedback coding on ($t \bmod c$). N_{GOP} is the total number of pictures in a GOP. S_k^t and $S_k^{t-N_{GOP}}$ represent the convolutional feature information (spatial representation) of kth CTU obtained from the original frame f_{org} at t position and reconstruction frame f_{rec} at $t - N_{GOP}$ position, respectively.

Specifically, if the encoding mode is intra coding, the spatial representation will directly input to the LB-PSO model. Otherwise, we compute the semantic residue information by applying the absolute difference between the current spatial representation S_k^t of the original CTU and the previous spatial representation $S_k^{t-N_{GOP}}$ of the reconstructed CTU before feeding it to the LB-PSO model to accurately generate rate control parameters on the changes between consecutive CTUs. In addition, the reconstructed frame at $t - N_{GOP}$ is chosen in the proposed method because a group of pictures in a video allows for exploits of the temporal redundancy in the video. The proposed model can be adapted following the N_{GOP}.

3.2. Learning-Based Particle Swarm Optimization Network
3.2.1. LB-PSO Estimator

Our LB-PSO is proposed to define the optimal mapping ϕ from the spatial–temporal representation of CTU g_k to rate control parameters y_k, $y_k = \{\alpha, \beta\}_k$. We introduce a feedforward network with one hidden layer to determine y_k. This feedforward network can be computed as in (8):

$$y_k = \phi(h_k; W_\phi, b_\phi) = W_\phi^T h_k + b_\phi \quad (8)$$

where W_ϕ provides the weights of a mapping function ϕ, b_ϕ is a bias, and h_k represents the output of the hidden layer. Precisely, h_k is designed by applying a rectified linear activation function to the output of a linear transformation composed of the weights W_h and bias b_h parameters to trigger a non-linear transformation. Thus, h_k can be derived as in (9):

$$h_k = \max\{0, W_h^T g_k + b_h\} \qquad (9)$$

From (8) and (9), our complete mapping model can be reformulated as in (10):

$$y_k = W_\phi^T \max\{0, W_h^T g_k + b_h\} + b_\phi \qquad (10)$$

The model parameters $M = \{W_\phi, W_h, b_\phi, b_h\}$ are optimized by utilizing swarm intelligence to exchange information between particles about R–D cost function, J. On the other hand, the model parameters regulate its trajectory concerning its best previous position and the best previous position reached by any member of its neighborhood. To target the swarm intelligence rule, the cost function J is determined by two objective functions, including a reconstruction error (MSE) of visual quality and smooth $_{L1}$ error of bit allocation. The cost function J can be defined as in (11):

$$J = \frac{1}{N}\sum_{j=0}^{N-1}(f_{org_j} - f_{rec_j})^2 + \eta \operatorname{smooth}_{L1}(R_T - R_A) \qquad (11)$$

$$\operatorname{smooth}_{L1}(U) = \begin{cases} \dfrac{U^2}{2}, & \text{if } |U| < 1 \\ |U| - \dfrac{1}{2}, & \text{otherwise} \end{cases} \qquad (12)$$

where N is the total number of pixels in a picture and η is a penalty coefficient. R_T and R_A are the target and actual bit on the picture level, respectively.

According to the cost function design, the model parameters are updated after all CTUs are fully encoded. This cost function is aimed at the model learning to achieve the trade-off between distortion and bit allocation. The next section introduces the complete process of the parameters update.

3.2.2. Parameter Updating

In this subsection, we present the parameter update of the encoder controller corresponding to the intra/inter coding mode. In addition, the inter coding mode is classified into two sets of coding frames: a core frame and a common one. A core frame is encoded by activating the re-feedback coding to adjust the bit budget at the CTU coding level. In contrast, the common frame is coded by applying the default Lagrangian multiplier to determine the bit budget at the CTU coding level. For both intra coding and core frame of inter coding, the bit budget at the CTU coding level is computed using Equations (4) and (10). Additionally, the model parameters M in Equation (10) individually parameterize its value according to its movement in a search space.

Let P be the total size of the population, V_i be the velocity (position change) of i-th particle, B_i be the best previous model parameters of i-th particle, and B_g be the best model parameter in the swarm. Then the swarm is manipulated on each iteration n according to the following two equations:

$$V_i^{n+1} = aV_i^n + c_1 r_{i1}^n (B_i^n - M_i^n) + c_2 r_{i2}^n (B_g^n - M_i^n), \qquad (13)$$

$$M_i^{n+1} = M_i^n + V_i^{n+1}, \qquad (14)$$

where $i = 1, 2, \ldots, P$, and a is the inertia weight of velocity V, which is used to control the trade-off between the swarm's global and local exploration capabilities. c_1 and c_2 are two positive acceleration constants, named the PSO's cognitive and social parameters, respectively. r_{i1} and r_{i2} are the random numbers, generated from a uniform distribution within the range $[0, 1]$. The performance of each model parameter M_i in the swarm is measured according to the cost function J. The lower cost function indicates a better M_i.

After finalizing the best M_i to preserve the minimal cost function J at the CTU coding level, the CTU is encoded.

For the picture level of inter coding, the rate control parameters are adjusted by considering the residue score of the semantic residue information. The probability of residue score Q^t on a picture at time t can be computed as

$$Q^t = \sum_{k \in K} \sum_{j \in S_k} \frac{A_k^t(j)}{S_k^t(j)} \tag{15}$$

$$A_k^t(j) = \begin{cases} 0, & \text{if } t - N_{GOP} \leq 0 \\ \left| S_k^t(j) - S_k^{t \times \lfloor \frac{t}{N_{GOP}} \rfloor}(j) \right|, & \text{otherwise} \end{cases} \tag{16}$$

where $\lfloor . \rfloor$ represents the rounded result. Additionally, in the GOP regarding the spatiotemporal information of the video sequence, the picture levels generally have different pairs of encoder controller coefficients α_p and β_p. Therefore, the rate control parameters can be updated by (17)–(21). The Lagrangian multiplier, λ, is defined as

$$\lambda = \alpha_{pold} \cdot R^{\beta_{pold}} \tag{17}$$

If the GOP_{id} equals 0, a pair of rate control parameters can be formulated as in (18) and (20).

$$\alpha_{pnew} = \alpha_{pold} + \delta_\alpha \cdot (\ln(\lambda_r - \lambda_c)) \cdot \alpha_{pold} + \zeta Q^t \tag{18}$$

$$\beta_{pnew} = \beta_{pold} + \delta_\beta \cdot (\ln(\lambda_r - \lambda_c)) \cdot \ln(bpp_r) + \frac{\zeta}{2} Q^t \tag{19}$$

Otherwise, a pair of rate control parameters can be computed as in (19) and (20).

$$\alpha_{pnew} = \alpha_{pold} + \zeta Q^t \tag{20}$$

$$\beta_{pnew} = \beta_{pold} + \frac{\zeta}{2} Q^t \tag{21}$$

where δ_α and δ_β are the default constant in HEVC reference software. λ_r represents the real λ value, λ_c is a computed λ value from the real cost bpp_r with the previous rate control parameters α_{pold} and β_{pold} at picture level, and ζ is the residue penalty constant.

For the quantization parameter (QP), it can be determined as in (21).

$$QP = 4.2005 \cdot \ln(\lambda) + 13.7122 \tag{22}$$

Figure 3 provides the model flowchart of the learning-based PSO method, named LB-PSO. LB-PSO initially randomizes the group of particle parameters. Then, the rate control coefficients are computed using the LB-PSO estimator. Subsequently, the LB-PSO model's best local and global parameters have reallocated if the current position is better than the stored position according to its cost function, J. After that, the velocity V and position M are calculated following Equations (13) and (14). Finally, the best particle for the LB-PSO model is determined to generate the best rate control coefficients for the current input CTU context.

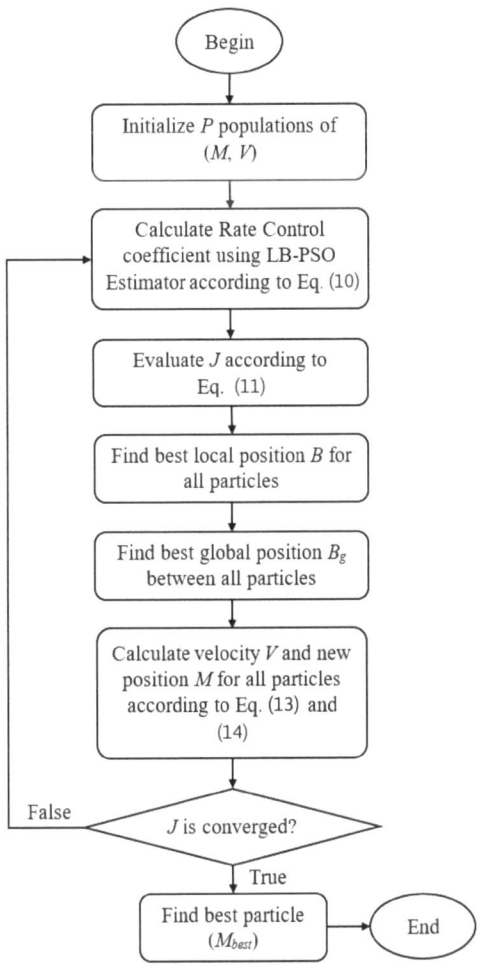

Figure 3. Learning-based particle swarm optimization flowchart.

4. Experimental Results

To evaluate the performance of the proposed learning-based particle swarm optimization, the experiments are conducted on various videos, including static and dynamic scenes.

4.1. Experiment Setting

In the experiment, the proposed algorithm is implemented on HEVC reference software [14] and is compared with the PS-GOP [36] and the state-of-the-art R–λ rate control (RC-HEVC) [11]. According to HEVC common parameter setting [3], the largest size of a CTU produces high-efficiency coding performance. Specifically, the largest feasible size of a CTU in HEVC is a 64 × 64 block size. We have also designed the model to adapt bit allocation for CTUs related to their spatial information, which is extracted using a pre-trained CNN model. Since we have implied CNN feature extraction on the largest size of a CTU in HEVC, we transform YUV420 format to a true color (64 × 64 × 3) CTU as the input in the feature extraction block. The proposed algorithm and baseline methods are simulated in the same reference software HM-16.10. Precisely, the experiments are conducted under the low-delay P main profile configurations, and the encoder parameters are set according to the standard setting in [35] by enabling the rate control as *True*. In addition, there are 100 iterations in every decision-making process for each rate control parameters prediction in the proposed LB-PSO. There are fifteen test video sequences with four video resolutions, including two videos of 240p (wide quarter video graphics array—WQVGA) [41], three videos of 480p (wide video graphics array—WVGA) [41], five videos of 720p (HD) [42], three videos of 1080p (full HD) [41], and two videos of 4k resolution [43]. Table 1 briefly summarizes the characteristics of the test video sequence. In addition, the test video sequence is encoded at four target bit rates corresponding to the video resolution. Since the goal of rate control is not only to improve the visual quality of the video for a given bit rate but also to achieve the bit rate closest to the target bit rate, both peak signal-to-noise ratio ($PSNR$) and bit rate error (BRE) are used as the criteria for determining the performance of the rate control algorithm. The $PSNR$ and BRE can be computed as in (22) and (23).

$$PSNR = 10\log\left[\frac{(2^n - 1)^2}{\frac{1}{N}\sum_{j=0}^{N-1}\left(f_{org_j} - f_{rec_j}\right)^2}\right] \quad (23)$$

$$BRE = \frac{(R_T - R_A)}{R_T} \times 100\% \quad (24)$$

where n represents bit depth.

Table 1. Characteristics of Test Video Sequences and Bit Rate.

Resolution	Name of Video Sequence	Frame Rate (fps)	Bit Rate (kbps)
3840 × 2160	HoneyBee	120	1000, 2000,
	Jocky	120	3000, 4000
1920 × 1080	ParkScene	24	
	Cactus	50	1000, 2000,
	BQTerrace	60	3000, 4000
1280 × 720	FourPeople	60	
	KristenAndSara	60	
	Vidyo1	60	384, 512,
	Vidyo3	60	850, 1200
	Vidyo4	60	
832 × 480	BasketballDrillText	50	
	PartyScene	50	384, 512,
	BQMall	60	768, 1200
416 × 240	BlowingBubbles	50	256, 384,
	BQSquare	60	512, 1200

4.2. Experimental Results and Analysis

(1) *R–D performance and Bit Rate Accuracy*: The first experiment was conducted on the low video resolution (WQVGA), which contains two video sequences with different frame rates, including BlowingBubbles and BQSquare. These two videos have various dynamic characteristics, such as a moving camera, moving objects, and illumination changes. Table 2 describes the proposed method's $PSNR$ and BRE performance compared with the baseline methods. Our learning-based method outperforms all the baseline methods as we achieve the highest $PSNR$ value with the same bit rate.

Table 2. The Performance of PSNR and BRE of Video Sequence with Resolution of 416 × 240.

Name of Video Sequence	Target Bit Rate	RC-HEVC			PS-GOP			Proposed Method		
		Bit Rate	PSNR	BRE	Bit Rate	PSNR	BRE	Bit Rate	PSNR	BRE
BlowingBubbles	256	256.06	29.69	−0.02	256.08	29.79	−0.03	256.02	29.99	−0.01
	384	384.05	31.14	−0.01	384.00	31.26	0.00	384.02	31.44	−0.01
	512	512.06	32.26	−0.01	512.05	32.38	−0.01	512.04	32.51	−0.01
	1200	1200.18	35.64	−0.02	1200.05	35.71	0.00	1200.15	35.73	−0.01
BQSquare	256	256.04	30.31	−0.02	256.01	30.42	−0.01	256.02	30.60	−0.01
	384	384.03	31.53	−0.01	384.03	31.67	−0.01	384.03	31.78	−0.01
	512	512.03	32.45	−0.01	512.03	32.56	−0.01	512.02	32.64	0.00
	1200	1200.06	35.20	0.00	1200.04	35.33	0.00	1200.04	35.37	0.00
	Average		32.28	−0.01		32.39	−0.01		**32.51**	−0.01

Specifically, our method's average *PSNR* enhancement is 0.23 dB and 0.12 dB compared with RC-HEVC and PS-GOP, respectively. Our approach also performs the maximum *PSNR* improvement (*max*) of 0.30 dB and 0.20 dB compared to RC-HEVC and PS-GOP. Figure 4a illustrates the *R–D* performance curve of the BQSquare test sequence. The learning-based approach obtains a better *R–D* performance than the baselines method. In addition, the average *BRE* of RC-HEVC, PS-GOP, and our methods are 0.01%, indicating that all approaches can effectively achieve the target bit rate. However, the proposed method has the lowest *BRE* at a lower target bit rate (256 kbps). It is noticed that the RC-HEVC has poor visual quality on these WQVGA with dynamic scenes compared to all approaches. As a result, even if the scene has dynamic properties, our algorithm can constructively achieve the target bit rate with the good visual quality of the WQVGA sequence.

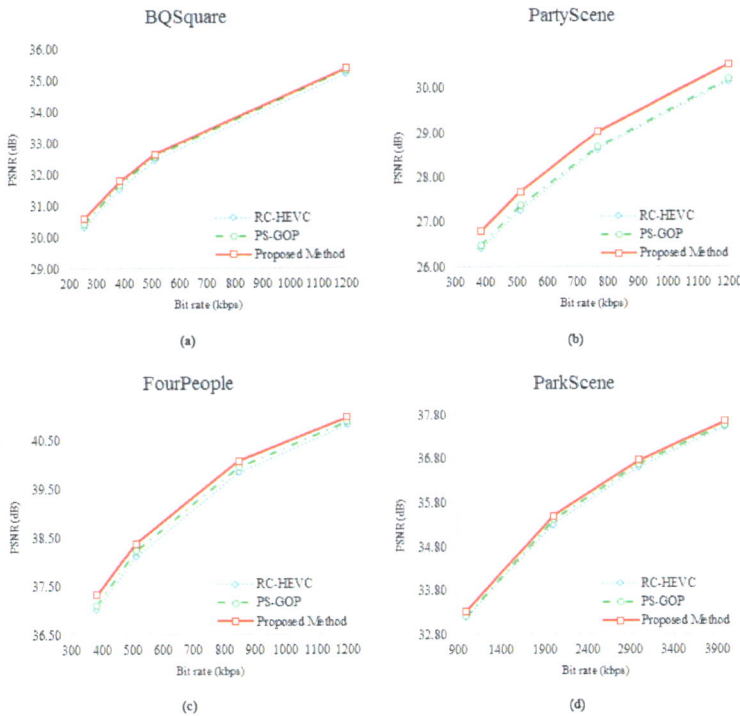

Figure 4. Rate–Distortion curves: (**a**) BQSquare, (**b**) PartyScene, (**c**) FourPeople, (**d**) ParkScene.

Next, the WVGA sequences were tested, such as BasketballDrillText, PartyScene, and BQMall. The scene properties are similar to the above experiments, but these WVGA sequences are more challenging than WQVGA because they involve multi-object movement, camera movement, and higher resolution. The outcomes of $PSNR$ and BRE are summarized in Table 3, where the proposed learning-based method works much better. It reaches 0.41 dB and 0.33 dB of visual quality better than RC-HEVC and PS-GOP, respectively. Concisely, our approach has no error bit consumption on average and performs 0.23 dB and 0.16 dB on average higher than RC-HEVC and PS-GOP, respectively. On one side of the R–D curve, our proposed method is significantly higher than the competitive methods, as shown in Figure 4b. Based on the outcomes of all approaches in Tables 2 and 3, the R–λ rate control and PS-GOP are unsuitable for such dynamic scenes and cameras. Consequently, it can indicate that the λ adjustment and quality control are not correctly estimated.

Table 3. The Performance of PSNR and BRE of Video Sequence with Resolution of 832 × 480.

Name of Video Sequence	Target Bit Rate	RC-HEVC			PS-GOP			Proposed Method		
		Bit Rate	PSNR	BRE	Bit Rate	PSNR	BRE	Bit Rate	PSNR	BRE
BasketballDrillText	384	384.03	30.82	−0.01	383.99	30.93	0.00	384.02	30.99	−0.01
	512	512.05	31.94	−0.01	512.00	32.01	0.00	511.99	32.08	0.00
	768	768.04	33.46	−0.01	768.04	33.52	−0.01	768.05	33.60	−0.01
	1200	1200.10	35.15	−0.01	1200.07	35.20	−0.01	1200.07	35.32	−0.01
PartyScene	384	384.01	26.40	0.00	384.00	26.49	0.00	383.97	26.80	0.01
	512	512.02	27.27	0.00	512.01	27.37	0.00	511.96	27.68	0.01
	768	768.09	28.61	−0.01	768.02	28.68	0.00	768.02	29.01	0.00
	1200	1200.06	30.15	−0.01	1200.02	30.20	0.00	1200.03	30.53	0.00
BQMall	384	384.01	30.68	0.00	384.13	30.77	−0.03	384.00	30.85	0.00
	512	512.01	31.86	0.00	512.05	31.92	−0.01	512.03	32.00	−0.01
	768	768.01	33.50	0.00	768.01	33.59	0.00	768.01	33.66	0.00
	1200	1200.04	35.28	0.00	1200.03	35.33	0.00	1200.01	35.39	0.00
Average			31.26	−0.01		31.33	−0.01		**31.49**	0.00

After testing the WVGA sequences, the HD videos containing video conferencing and online teaching test sequences were simulated. The HD videos are FourPeople, KristenAndSara, Vidyo1, Vidyo3, and Vidyo4. These videos have the characteristics of a static camera with multiple objects moving. Figure 4c shows an overall outgrowth of the R–D curve of FourPeople from the low bit rate to the high bit rate. Although the scene is used with a static camera, the proposed method's R–D performance is noticeably greater than the competitive methods. Additionally, the $PSNR$ and BRE evaluations of these HD video sequences are recorded in Table 4, where the average $PSNR$ enhancement value of our method is approximately 0.17 dB (max = 0.30 dB) and 0.08 dB (max = 0.21 dB) in comparison with the RC-HEVC and PS-GOP.

Table 4. The Performance of PSNR and BRE of Video Sequence with Resolution of 1280 × 720.

Name of Video Sequence	Target Bit Rate	RC-HEVC			PS-GOP			Proposed Method		
		Bit Rate	PSNR	BRE	Bit Rate	PSNR	BRE	Bit Rate	PSNR	BRE
FourPeople	384	383.97	37.02	0.01	383.99	37.12	0.00	383.99	37.32	0.00
	512	511.97	38.10	0.01	512.00	38.24	0.00	511.99	38.38	0.00
	850	849.98	39.84	0.00	849.99	39.94	0.00	849.98	40.06	0.00
	1200	1200.08	40.81	−0.01	1199.96	40.87	0.00	1200.05	40.97	0.00
KristenAndSara	384	384.06	39.17	−0.02	384.08	39.32	−0.02	384.12	39.37	−0.03
	512	512.07	40.03	−0.01	512.09	40.17	−0.02	512.11	40.20	−0.02
	850	850.12	41.31	−0.01	850.09	41.43	−0.01	850.12	41.47	−0.01
	1200	1200.18	42.04	−0.01	1200.16	42.12	−0.01	1200.16	42.16	−0.01
Vidyo1	384	384.00	38.95	0.00	383.98	39.06	0.01	384.00	39.11	0.00
	512	512.01	39.86	0.00	511.93	39.95	0.01	511.99	40.01	0.00
	850	849.96	41.19	0.00	849.88	41.26	0.01	850.01	41.32	0.00
	1200	1200.00	41.93	0.00	1199.96	42.00	0.00	1200.01	42.07	0.00
Vidyo3	384	384.01	37.85	0.00	384.00	38.00	0.00	384.02	38.01	−0.01
	512	512.02	38.82	0.00	512.01	38.95	0.00	512.01	38.97	0.00
	850	850.01	40.22	0.00	850.01	40.33	0.00	850.01	40.37	0.00
	1200	1200.02	41.00	0.00	1200.03	41.08	0.00	1200.00	41.12	0.00
Vidyo4	384	384.01	38.68	0.00	384.01	38.73	0.00	384.01	38.86	0.00
	512	512.02	39.47	0.00	512.01	39.53	0.00	512.02	39.67	0.00
	850	850.02	40.67	0.00	850.01	40.74	0.00	850.02	40.86	0.00
	1200	1200.02	41.39	0.00	1200.05	41.45	0.00	1200.02	41.54	0.00
	Average		39.92	0.00		40.02	0.00		**40.09**	0.00

The last experiment was applied on full HD and 4k video test sequences. The first three videos, ParkScene, Cactus, and BQTerrace, were used for the full HD experiment. The last two sequences, HoneyBee and Jocky, were used for 4k videos. This last test contained all types of scenarios. The ParkScene and Jocky videos have a moving camera and multiple object motions, while the BQTerrace video stacks the camera motion with a static camera. Furthermore, the Cactus video consists of a static camera and the rotation of the objects. The HoneyBee video has multiple object motions and a static camera. According to Table 5, the overall $PSNR$ evaluation of the proposed method on the BQTerrace sequence at a low bit rate is the highest compared to the other sequences. In contrast, the ParkScene sequence has the highest $PSNR$ at a high bit rate. The reason is that the scenes containing a dynamic camera have significant movement changes; thus, the state-of-the-art R–λ rate control cannot update the encoding controller correctly. In addition, PS-GOP uses parameter sharing in GOP, which is not enough to adapt to encoder parameters following frame characteristics. Reasoning from this fact, our method establishes a novel mapping between frame features and R–λ coefficient parameters. We provide a computationally feasible solution using LB-PSO to produce optimal R–D for good visual quality and to maintain the target bit rate. Figure 4 shows the overall R–D curve on different video resolutions. Consequently, our method has achieved the highest outcomes of all competitive methods. From Table 2 to Table 5, the average $PSNR$ improvement is 0.19 dB (max = 0.41 dB) and 0.10 dB (max = 0.33 dB) compared with RC-HEVC and PS-GOP, respectively.

Table 5. The Performance of PSNR and BRE of Video Sequence with Resolution of 1920 × 1080 and 4k.

Name of Video Sequence	Target Bit Rate	RC-HEVC			PS-GOP			Proposed Method		
		Bit Rate	PSNR	BRE	Bit Rate	PSNR	BRE	Bit Rate	PSNR	BRE
ParkScene	1000	999.96	33.20	0.00	999.84	33.21	0.02	999.86	33.32	0.01
	2000	2000.01	35.30	0.00	1999.89	35.41	0.01	2000.10	35.49	0.00
	3000	2999.95	36.60	0.00	2999.91	36.68	0.00	2999.98	36.76	0.00
	4000	4000.11	37.52	0.00	4000.09	37.57	0.00	4000.11	37.66	0.00
Cactus	1000	1000.01	31.62	0.00	1000.02	31.75	0.00	1000.02	31.74	0.00
	2000	2000.04	33.77	0.00	2000.03	33.85	0.00	2000.03	33.87	0.00
	3000	3000.09	34.96	0.00	3000.03	35.01	0.00	3000.03	35.04	0.00
	4000	4000.06	35.70	0.00	3999.95	35.77	0.00	4000.07	35.81	0.00
BQTerrace	1000	1000.05	31.62	−0.01	1000.01	31.73	0.00	1000.17	31.97	−0.02
	2000	2000.13	33.03	−0.01	2000.02	33.11	0.00	2000.04	33.25	0.00
	3000	3000.15	33.67	0.00	3000.01	33.78	0.00	3000.08	33.82	0.00
	4000	4000.53	34.10	−0.01	4000.05	34.20	0.00	4000.04	34.15	0.00
HoneyBee	1000	1000.01	38.24	0.00	1000.00	38.25	0.00	1000.03	38.31	0.00
	2000	2000.01	38.63	0.00	2000.00	38.65	0.00	2000.01	38.66	0.00
	3000	3000.01	38.75	0.00	3000.01	38.78	0.00	3000.01	38.78	0.00
	4000	4000.40	38.81	−0.01	4000.01	38.83	0.00	4000.02	38.83	0.00
Jocky	1000	999.98	32.30	0.00	1000.01	32.40	0.00	1000.00	32.40	0.00
	2000	2000.03	35.55	0.00	2000.01	35.60	0.00	2000.00	35.61	0.00
	3000	3000.00	36.95	0.00	3000.04	36.97	0.00	3000.06	36.99	0.00
	4000	4000.00	37.68	0.00	3999.99	37.69	0.00	4000.02	37.71	0.00
Average			35.40	0.00		35.46	0.00		**35.51**	0.00

The PSNR performance of our proposed model is extensively compared with other state-of-the-art rate control methods for both the dynamic scene and interview scene as shown in Table 6. Our proposed model achieves the highest PSNR for all bit rates in both types of video sequences. This indicates that the inter coding approach should not only consider the inter-block dependency coding structure but also the rate control coefficient.

Table 6. PSNR Comparisons at different bit rates with other state-of-the-art rate control schemes.

Name of Video Sequence	Bit Rate	BA [44]	BAF [45]	RCA [37]	Proposed Method
FourPeople	384	36.30	36.81	37.07	**37.32**
	512	37.49	38.19	38.31	**38.38**
	850	39.76	39.98	40.03	**40.06**
	1200	40.52	40.69	40.89	**40.97**
BasketballDrillText	384	30.82	30.81	30.89	**30.99**
	512	31.87	31.86	31.91	**32.08**
	850	33.41	33.44	33.52	**33.60**
	1200	34.91	34.96	35.19	**35.32**

Additionally, Figure 5 shows a graph of the PSNR difference between consecutive frames. The plot shows that the performance of the proposed method adaptively achieves better results on frame reconstruction from the start of encoding compared to RC-HEVC and PS-GOP. This demonstrates the effective interaction of spatiotemporal features in the rate control model and the crossed LB-PSO model to decide on appropriate rate control coefficients to acquire the target bit rate and perform well in PSNR. Furthermore, Figure 6 indicates the details of the rate fluctuation performance of the proposed method compared to the baselines. This rate fluctuation describes successive frames' historical bit allocation performance to understand the bit flow in the video codec. Therefore, LB-PSO can control bit allocation better than the baselines, and it can carry out lower bit allocation and produce higher PSNR in most consecutive frames, as shown in Figures 5 and 6.

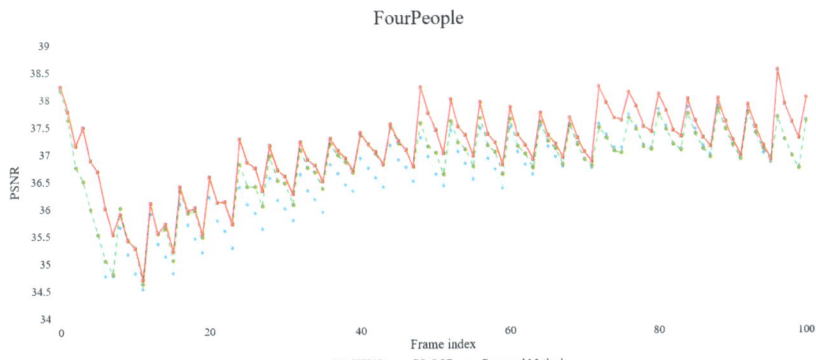

Figure 5. Comparison of PSNR difference between consecutive frames.

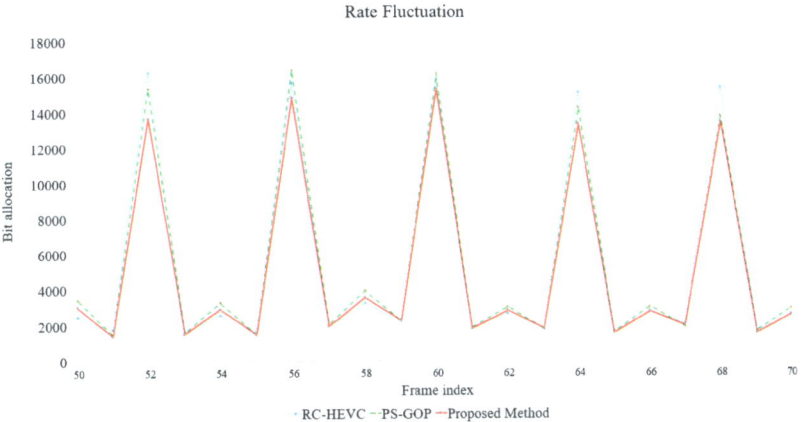

Figure 6. Rate fluctuation performance comparison.

(2) *Bit Heatmaps and Visual Quality*: To indicate the performance of bit allocation at the CTU level, the heatmap visualization and the subjective results of the reconstructed frame are illustrated in Figures 7 and 8. Since there is no modification of the intra coding of PS-GOP, Figure 7 shows only the comparison between state-of-the-art RC-HEVC with our proposed learning-based approach. The bit consumption is highlighted by red color intensity on each CTU, while the blue acts as a mask to cover the frame. If the red intensity is low, the allocated bits are consumed less. The patch image is extracted from the frame to illustrate the greatest difference in bit consumption at the CTU level of RC-HEVC and our proposed method. Figure 7b,c reveal that the bit allocation performance of RC-HEVC on the plane space CTU is slightly high, leading to less bit budget for the necessary spatial CTU. On the contrary, our proposed method obtains smoother bit allocation on non-important spatial images (low-frequency components), providing more budget to important CTU features. Additionally, the visualization of the human face of the proposed learning-based approach on the intra-picture shows more details with a smoother look than that of RC-HEVC, as shown in the green box of Figure 7c,d. According to these results, our LB-PSO can obtain better bit allocation by using the information from the mapping encoder control parameters with the input convolution feature map of each spatial CTU instead of the fixed initialization of R–λ rate control.

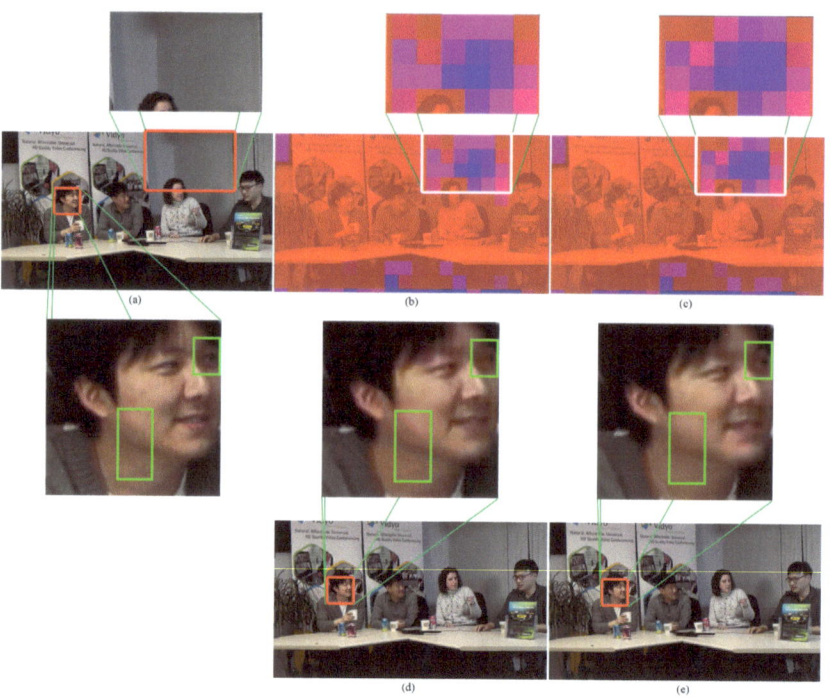

Figure 7. Bit heatmaps and reconstructed frame of intra coding at 384 kbps: (**a**) original frame, (**b**,**d**) RC-HEVC, and (**c**,**e**) proposed method.

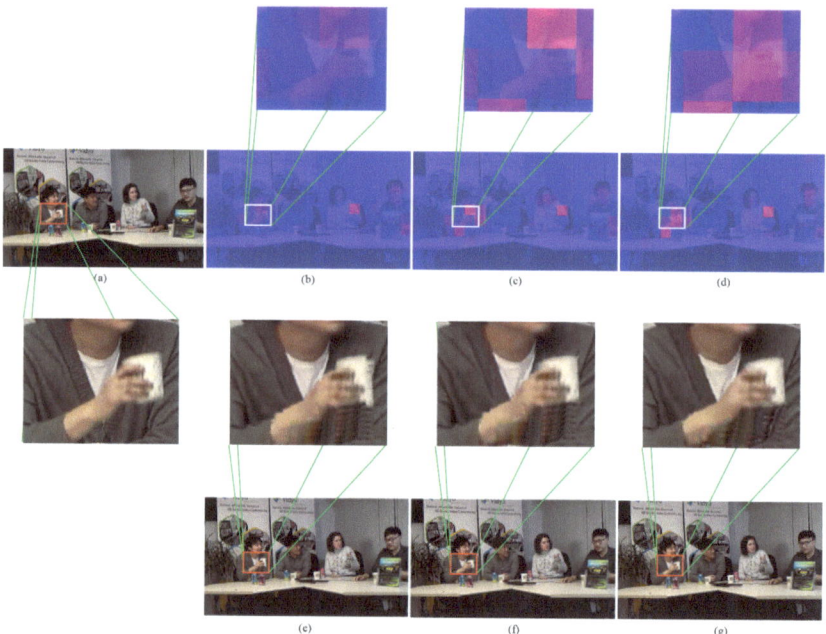

Figure 8. Bit heatmaps and reconstructed frame of inter coding at 384 kbps: (**a**) original frame, (**b**,**e**) RC-HEVC, (**c**,**f**) PS-GOP, and (**d**,**g**) proposed method.

For inter coding, the PS-GOP is added in comparison. Similarly, the color representation is defined the same as the intra coding. Regarding bitmaps, Figure 8b shows that RC-HEVC has a problem with bit allocation on the essential features. Due to hand movement, RC-HEVC should provide higher bit allocation in these necessary parts; on the contrary, it allocates fewer bits to these blocks. Furthermore, PS-GOP attempts to allocate the amount of bit budget to the hand movement area to keep the visual quality of the action consistent. However, the bit budget on large hand motion blocks is still small, as shown in Figure 8c.

Regarding residual semantic information, our proposed method can correctly regulate the bit budget responding to the motion information in the scene, as illustrated in Figure 8d. On the other hand, our proposed method obtains the accurate bit allocation of each CTU corresponding to its spatial–temporal characteristics. Furthermore, the visual quality visualization of this hand movement is shown in Figure 8e–g. In particular, RC-HEVC has a considerable distortion in this hand movement area, while PS-GOP is slightly better than RC-HEVC. Although PS-GOP is better than RC-HEVC, PS-GOP still has higher distortion compared with our proposed method. As a result, the proposed method achieves better hand and cup shapes than the competitive methods. According to our experimental results, we can conclude that the proposed learning-based R–λ parameter outperforms other competing methods by achieving the highest $PSNR$ while maintaining the target bit rate.

(3) *Computational Complexity*: We compare the computational time of the proposed method with RC-HEVC and PS-GOP. Regarding computational time in an average of seconds per frame, as indicated in Table 7, our LB-PSO achieves 53.30 s/frame, 97.79 s/frame, and 351.10 s/frame on WVGA, HD, and full HD resolution, respectively. We also compare our computational complexity with other baseline methods. Table 6 shows that our computational time is higher than the baseline methods. This is because our framework is designed as online training using the integration of the forward pass network with particle swarm optimization. However, we obtained a significantly higher PSNR value and achieved the target bit rate. Furthermore, our bit allocation was assigned correctly compared to baseline approaches.

Table 7. Computational Complexity.

Intel Core i9-7960× CPU @ 2.80 GHz			
Resolution	HM-16.10 (s/frame)	PS-GOP (s/frame)	Proposed Method (s/frame)
WVGA	24.10	23.75	53.30
HD	45.18	44.92	97.79
Full HD	166.15	165.47	351.10
Average	78.48	78.04	167.40

5. Conclusions

In this paper, we proposed novel learning-based R–λ parameters for HEVC. The proposed framework is embedded with a deep convolution neural network feature map and LB-PSO, which brings advantages to rate control parameters estimation corresponding to spatial–temporal CTUs. LB-PSO is designed to obtain the feasible rate control coefficient parameters solution to optimize the R–D relationship. Experimental results clearly show that our proposed learning-based approach obtains an accurate target bit rate with 0.19 dB on average to 0.41 dB and 0.10 dB on average to 0.33 dB maximum $PSNR$ improvement than the state-of-the-art RC-HEVC and PS-GOP, accordingly. Due to the bit allocation, our algorithm can achieve an operational bit distribution to each CTU on both intra and inter coding. In other words, our method is effective and robust for determining the bit budget for the CTU of the frame. For future work, CTU partitioning will be considered together with R–λ parameters to increase coding efficiency.

Author Contributions: Conceptualization, S.A.; Methodology, S.C.; Software, S.C.; Validation, S.A.; Formal analysis, S.C.; Investigation, S.C.; Resources, S.A.; Data curation, S.C.; Writing—original draft, S.C.; Writing—review & editing, S.A. and Y.M.; Visualization, S.C.; Supervision, S.A. and Y.M.; Project administration, S.A.; Funding acquisition, S.A. All authors have read and agreed to the published version of the manuscript.

Funding: This research work is supported by the Collaborative Research Project entitled Video Processing and Transmission by the JICA Project for AUN/SEED-Net, Japan, Thailand Science research and Innovation Fund Chulalongkorn University (CU_FRB65_ind (9)_157_21_23), (IND66210019), the NSRF via the Program Management Unit for Human Resources Institutional Development, Research and Innovation [grant number B04G640053], and Ratchadaphiseksomphot Endowment Fund (Multimedia Data Analytics and Processing Research Unit).

Institutional Review Board Statement: Not applicable.

Informed Consent Statement: Not applicable.

Data Availability Statement: Not applicable.

Conflicts of Interest: The authors declare no conflict of interest.

References

1. Owens, J. *Television Production*; CRC Press: Boca Raton, FL, USA, 2015.
2. Cisco. Cisco Annual Internet Report—Cisco Annual Internet Report (2018–2023) White Paper. 9 March 2020. Available online: https://www.cisco.com/c/en/us/solutions/collateral/executive-perspectives/annual-internet-report/white-paper-c11-741490.html (accessed on 11 Feburary 2021).
3. Wiegand, T.; Sullivan, G.J.; Bjontegaard, G.; Luthra, A. Overview of the H. 264/AVC video coding standard. *IEEE Trans. Circuits Syst. Video Technol.* **2003**, *13*, 560–576. [CrossRef]
4. Sullivan, G.J.; Ohm, J.-R.; Han, W.-J.; Wiegand, T. Overview of the high efficiency video coding (HEVC) standard. *IEEE Trans. Circuits Syst. Video Technol.* **2012**, *22*, 1649–1668. [CrossRef]
5. Correa, G.; Assuncao, P.; Agostini, L.; da Silva Cruz, L.A. Performance and computational complexity assessment of high-efficiency video encoders. *IEEE Trans. Circuits Syst. Video Technol.* **2012**, *22*, 1899–1909. [CrossRef]
6. Grois, D.; Marpe, D.; Mulayoff, A.; Itzhaky, B.; Hadar, O. Performance comparison of H. 265/MPEG-HEVC, vp9, and H. 264/MPEG-AVC encoders. In Proceedings of the 2013 Picture Coding Symposium (PCS), San Jose, CA, USA, 8–11 December 2013; pp. 394–397.
7. Sze, V.; Budagavi, M.; Sullivan, G.J. High efficiency video coding (HEVC). In *Integrated Circuit and Systems, Algorithms and Architectures*; Springer: Berlin/Heidelberg, Germany, 2014; Volume 39, p. 40.
8. Lainema, J.; Bossen, F.; Han, W.-J.; Min, J.; Ugur, K. Intra coding of the HEVC standard. *IEEE Trans. Circuits Syst. Video Technol.* **2012**, *22*, 1792–1801. [CrossRef]
9. Fu, C.M.; Alshina, E.; Alshin, A.; Huang, Y.W.; Chen, C.Y.; Tsai, C.Y.; Hsu, C.W.; Lei, S.M.; Park, J.H.; Han, W.J. Sample adaptive offset in the HEVC standard. *IEEE Trans. Circuits Syst. Video Technol.* **2012**, *22*, 1755–1764. [CrossRef]
10. Lv, H.; Wang, R.; Xie, X.; Jia, H.; Gao, W. A comparison of fractional-pel interpolation filters in HEVC and H. 264/AVC. In Proceedings of the 2012 Visual Communications and Image Processing, San Diego, CA, USA, 27–30 November 2012; pp. 1–6.
11. Lv, H.; Wang, R.; Xie, X.; Jia, H.; Gao, W. λ domain rate control algorithm for High Efficiency Video Coding. *IEEE Trans. Image Process.* **2014**, *23*, 3841–3854.
12. He, Z.; Kim, Y.K.; Mitra, S.K. Low-delay rate control for DCT video coding via ρ-domain source modeling. *IEEE Trans. Circuits Syst. Video Technol.* **2001**, *11*, 928–940.
13. Liu, M.; Guo, Y.; Li, H.; Chen, C.W. Low-complexity rate control based on ρ-domain model for scalable video coding. In Proceedings of the 2010 IEEE International Conference on Image Processing, Hong Kong, China, 26–29 September 2010; pp. 1277–1280.
14. HEVC Reference Software. 2014. Available online: http://hevc.kw.bbc.co.uk/trac/browser/jctvc-hm/tags (accessed on 16 February 2023).
15. He, K.; Zhang, X.; Ren, S.; Sun, J. Deep residual learning for image recognition. In Proceedings of the IEEE Conference on Computer Vision and Pattern Recognition, Las Vegas, NV, USA, 27–30 June 2016; pp. 770–778.
16. Deng, J.; Dong, W.; Socher, R.; Li, L.-J.; Li, K.; Fei-Fei, L. Imagenet: A large-scale hierarchical image database. In Proceedings of the 2009 IEEE Conference on Computer Vision and Pattern Recognition, Miami, FL, USA, 20–25 June 2009; pp. 248–255.
17. Mohamed, A.W.; Sabry, H.Z. Constrained optimization based on modified differential evolution algorithm. *Inf. Sci.* **2012**, *194*, 171–208. [CrossRef]
18. Bertsekas, D.P. *Constrained Optimization and Lagrange Multiplier Methods*; Academic Press: Cambridge, MA, USA, 2014.
19. Beyer, H.-G.; Schwefel, H.-P. Evolution strategies—A comprehensive introduction. *Nat. Comput.* **2002**, *1*, 3–52. [CrossRef]
20. Dowsland, K.A.; Thompson, J. Simulated annealing. In *Handbook of Natural Computing*; Spring: 2012 ; pp. 1623–1655.

21. Mirjalili, S. Genetic algorithm. In *Evolutionary Algorithms and Neural Networks*: Springer: Berlin/Heidelberg, Germany, 2019; pp. 43–55.
22. Parsopoulos, K.E.; Vrahatis, M.N. Recent approaches to global optimization problems through particle swarm optimization. *Nat. Comput.* **2002**, *1*, 235–306. [CrossRef]
23. Du, K.-L.; Swamy, M. Particle swarm optimization. In *Search and Optimization by Metaheuristics*; Springer: Berlin/Heidelberg, Germany, 2016; pp. 153–173.
24. Tang, B.; Zhu, Z.; Luo, J. A framework for constrained optimization problems based on a modified particle swarm optimization. *Math. Probl. Eng.* **2016**, *2016*, 8627083. [CrossRef]
25. Khalili-Damghani, K.; Abtahi, A.-R.; Tavana, M. A new multi-objective particle swarm optimization method for solving reliability redundancy allocation problems. *Reliab. Eng. Syst. Saf.* **2013**, *111*, 58–75. [CrossRef]
26. Zhang, Y.; Wu, L.; Wang, S. UCAV path planning by fitness-scaling adaptive chaotic particle swarm optimization. *Math. Probl. Eng.* **2013**, *2013*, 705238. [CrossRef]
27. Xu, L.; Wang, J.; Li, Y.-P.; Li, Q.; Zhang, X. Resource allocation algorithm based on hybrid particle swarm optimization for multiuser cognitive OFDM network. *Expert Syst. Appl.* **2015**, *42*, 7186–7194. [CrossRef]
28. Darwish, A.; Ezzat, D.; Hassanien, A.E. An optimized model based on convolutional neural networks and orthogonal learning particle swarm optimization algorithm for plant diseases diagnosis. *Swarm Evol. Comput.* **2020**, *52*, 100616. [CrossRef]
29. Choi, H.; Yoo, J.; Nam, J.; Sim, D.; Bajić, I.V. Pixel-wise unified rate-quantization model for multi-level rate control. *IEEE J. Sel. Top. Signal Process.* **2013**, *7*, 1112–1123. [CrossRef]
30. Hu, S.; Wang, H.; Kwong, S. Adaptive quantization-parameter clip scheme for smooth quality in H. 264/AVC. *IEEE Trans. Image Process.* **2011**, *21*, 1911–1919.
31. Wang, S.; Ma, S.; Wang, S.; Zhao, D.; Gao, W. Quadratic ρ-domain based rate control algorithm for HEVC. In Proceedings of the 2013 IEEE International Conference on Acoustics, Speech and Signal Processing, Vancouver, BC, Canada, 26–31 May 2013; pp. 1695–1699.
32. Wang, S.; Ma, S.; Wang, S.; Zhao, D.; Gao, W. Rate-GOP based rate control for high efficiency video coding. *IEEE J. Sel. Top. Signal Process.* **2013**, *7*, 1101–1111. [CrossRef]
33. Sanz-Rodríguez, S.; Schierl, T. A rate control algorithm for HEVC with hierarchical GOP structures. In Proceedings of the 2013 IEEE International Conference on Acoustics, Speech and Signal Processing, Vancouver, BC, Canada, 26–31 May 2013; pp. 1719–1723.
34. Wang, M.; Ngan, K.N.; Li, H. Low-delay rate control for consistent quality using distortion-based Lagrange multiplier. *IEEE Trans. Image Process.* **2016**, *25*, 2943–2955. [CrossRef]
35. Bossen, F. Common Test Conditions and Software Reference Configurations. *JCTVC-J1100* **2012**, *12* .
36. Chen, S.; Aramvith, S.; Miyanaga, Y. Encoder Control Enhancement in HEVC Based on R-Lambda Coefficient Distribution. In Proceedings of the 2019 International Symposium on Multimedia and Communication Technology (ISMAC), Quezon City, Philippines, 19–21 August 2019; pp. 1–4.
37. Guo, H.; Zhu, C.; Xu, M.; Li, S. Inter-block dependency-based CTU level rate control for HEVC. *IEEE Trans. Broadcast.* **2019**, *66*, 113–126. [CrossRef]
38. Zhou, M.; Wei, X.; Kwong, S.; Jia, W.; Fang, B. Rate control method based on deep reinforcement learning for dynamic video sequences in HEVC. *IEEE Trans. Multimed.* **2020**, *23*, 1106–1121. [CrossRef]
39. Marzuki, I.; Lee, J.; Wiratama, W.; Sim, D. Deep convolutional feature-driven rate control for the HEVC encoders. *IEEE Access* **2021**, *9*, 162018–162034. [CrossRef]
40. Guo, Y.; Liu, Y.; Oerlemans, A.; Lao, S.; Wu, S.; Lew, M.S. Deep learning for visual understanding: A review. *Neurocomputing* **2016**, *187*, 27–48. [CrossRef]
41. ISO/IEC-JCT1/SC29/WG11, *Common Test Conditions of 3DV Core Experiments*; ISO: San Jose, CA, USA , 2014
42. Xiph.org. Xiph.org Video Test Media. 2017. Available online: https://media.xiph.org/video/derf/ (accessed on 16 February 2023).
43. Mercat, A.; Viitanen, M.; Vanne, J. UVG dataset: 50/120fps 4K sequences for video codec analysis and development. In Proceedings of the 11th ACM Multimedia Systems Conference, Istanbul, Turkey, 8–11 June 2020. Available online: https://ultravideo.fi/#testsequences (accessed on 16 February 2023).
44. Li, L.; Li, B.; Li, H.; Chen, C.W. λ-domain optimal bit allocation algorithm for High Efficiency Video Coding. *IEEE Trans. Circuits Syst. Video Technol.* **2016**, *28*, 130–142. [CrossRef]
45. Guo, H.; Zhu, C.; Li, S.; Gao, Y. Optimal bit allocation at frame level for rate control in HEVC. *IEEE Trans. Broadcast.* **2018**, *65*, 270–281. [CrossRef]

Disclaimer/Publisher's Note: The statements, opinions and data contained in all publications are solely those of the individual author(s) and contributor(s) and not of MDPI and/or the editor(s). MDPI and/or the editor(s) disclaim responsibility for any injury to people or property resulting from any ideas, methods, instructions or products referred to in the content.

Article

A Highly Pipelined and Highly Parallel VLSI Architecture of CABAC Encoder for UHDTV Applications

Chen Fu [1], Heming Sun [2], Zhiqiang Zhang [1] and Jinjia Zhou [1,*]

[1] Graduate School of Science and Engineering, Hosei University, Tokyo 184-8584, Japan
[2] Waseda Research Institute for Science and Engineering, Waseda University, Tokyo 169-8050, Japan
* Correspondence: zhou@hosei.ac.jp

Abstract: Recently, specifically designed video codecs have been preferred due to the expansion of video data in Internet of Things (IoT) devices. Context Adaptive Binary Arithmetic Coding (CABAC) is the entropy coding module widely used in recent video coding standards such as HEVC/H.265 and VVC/H.266. CABAC is a well known throughput bottleneck due to its strong data dependencies. Because the required context model of the current bin often depends on the results of the previous bin, the context model cannot be prefetched early enough and then results in pipeline stalls. To solve this problem, we propose a prediction-based context model prefetching strategy, effectively eliminating the clock consumption of the contextual model for accessing data in memory. Moreover, we offer multi-result context model update (MCMU) to reduce the critical path delay of context model updates in multi-bin/clock architecture. Furthermore, we apply pre-range update and pre-renormalize techniques to reduce the multiplex BAE's route delay due to the incomplete reliance on the encoding process. Moreover, to further speed up the processing, we propose to process four regular and several bypass bins in parallel with a variable bypass bin incorporation (VBBI) technique. Finally, a quad-loop cache is developed to improve the compatibility of data interactions between the entropy encoder and other video encoder modules. As a result, the pipeline architecture based on the context model prefetching strategy can remove up to 45.66% of the coding time due to stalls of the regular bin, and the parallel architecture can also save 29.25% of the coding time due to model update on average under the condition that the Quantization Parameter (QP) is equal to 22. At the same time, the throughput of our proposed parallel architecture can reach 2191 Mbin/s, which is sufficient to meet the requirements of 8 K Ultra High Definition Television (UHDTV). Additionally, the hardware efficiency (Mbins/s per k gates) of the proposed architecture is higher than that of existing advanced pipeline and parallel architectures.

Keywords: high efficiency video coding (HEVC); entropy coding; context adaptive binary arithmetic coding (CABAC); video coding; hardware design

1. Introduction

The creation of intelligent sensor nodes that enable intelligent processing for Internet of Things (IoT) surveillance, remote sensing, and smart city applications is gaining more and more attention [1]. In this, video data is crucial, and specifically designed video codecs have been preferred in recent years [2]. With a focus on reducing the data burden and improving the video quality [3], video coding and processing techniques performed in low-cost implementations and higher compression efficiency will cope with the design requirements of sensor nodes. The Joint Collaborative Team on Video Coding (JCT-VC) published the High Efficiency Video Coding (HEVC) standard in 2013 [4]. With a more flexible block division structure, a more precise coding mode, and some cutting-edge coding tools, HEVC is the widely used worldwide video coding standard [5].

The HEVC standard's coding structure primarily comprises Prediction, Estimation, Motion compensation, Quantization and Transform, and Entropy coding. The video

pixel value, which is broken down into two chrominance channels and one brightness channel, serves as the input for this coding system. The image is chunked into coding tree units (CTUs), which support a range of sizes [6]. Intra and inter frame prediction is first carried out to encode this CTU video block [7]. The rate-distortion cost is then assessed using various prediction modes, block size, and distortion degree, and the block segmentation method and the prediction mode of this CTU are obtained [8]. To produce the converted coefficient known as "residual", the original pixel data must be subtracted from the predicted data in order to obtain the difference [9]. The difference is then further transformed and quantized [10], and some high-frequency components are removed. To create the code stream, the projected data and the residuals will be entropy coded.

Entropy coding is a coding class that performs lossless coding based on the information entropy principle. Contrary to the Context-based Adaptive Variable Length Coding (CAVLC) and CABAC hybrid coding approach employed by the previous generation video coding standard AVC/H.264 [11], HEVC/H.265 only uses the CABAC entropy coding method.

A high-performance entropy encoder remains one of the hardware implementations' constraints for entropy coding in video coding. The amount of data that must be processed via entropy coding in HEVC is also significantly increased to handle more complex Rate Distortion Optimization (RDO) operations and Syntax Elements (SEs), which places more demands on hardware implementation. Parallel processing is challenging to implement because of the stringent data reliance of the binary arithmetic coding employed in CABAC, as well as the complexity of the arithmetic coding procedures, which might make it challenging to increase the primary frequency [12,13]. Ding et al. [14] proposed an optimized CABAC "Producer–Consumer" architecture through data flow modeling to achieve high throughput and low resource consumption. Wahiba et al. [15] proposed the processing of 1 to 5 bypass bins at the same by duplicating the number of bypass encoding engine (BEE) blocks for improving the throughput to be transmitted or stored. Ramos et al. [16] presented a novel scheme for multiple bypass bin processing, named multiple bypass bin scheme (MBBS), and the proposed method application into a baseline binary arithmetic encoding (BAE) architecture, showing an increasing bin per cycle throughput. Li et al. [17] considered the bypass mode encoding process in the CABAC and tried to merge bypass bins, and implemented one clock to encode six bins in bypass encoding mode to improve throughput. Zhou et al. [18] proposed and implemented in hardware a series of throughput improvement techniques: pre-normalization, Hybrid Path Coverage, Lookahead rLPS, bypass bin splitting and State Dual Transition, and by combining all these optimizations, overall CABAC performance improved by leaps and bounds.

The throughput rates of the SE generation and processing module and the BAE module are essential because they are two modules that both supply and process data. Consequently, we must address the latency that the complex data preparation required by the higher-level modules results in. Wahiba et al. [19] propose a new Register Transfer Level (RTL) architecture of HEVC CABAC encoder, where all SEs transmitted for 4×4 sub-blocks are studied and implemented. Saggiorato et al. [20] propose a novel efficient multi-core architectural approach, named Multiple Residual Syntax Element Treatment (MRSET), to meet the requirements of these recent CABAC designs. Tran et al. [21] and Nagaraju et al. [22] propose efficient hardware implementations of binarization for CABAC that focus on low area cost and power consumption while providing enough bins for high-throughput CABAC.

There is a problem that they need to address specifically, even though the current work considerably increases the throughput of CABAC encoders. When encoding successive bins of the same context model in BAE, the pipeline or parallel architecture of CABAC periodically stall, decreasing the coding efficiency. This paper aims to improve the performance further and enhance the compatibility of the entropy coding module, which is used to ensure the overall video coding architecture and the continuous and stable operation of this entropy coding encoder. This study builds on our earlier work by offering several fresh architectural modifications to enhance the critical path delay and the number of bins provided every clock cycle, dramatically increasing the overall throughput. Below is a summary of this paper's significant contributions.

1. We examine the challenges and bottlenecks in pipelined or parallel implementations brought on by arithmetic coding's back-and-forth dependency on coding states. We propose to use pre-range update and pre-renormalize technique to reduce the multiplex BAE route delay due to the incomplete reliance of the encoding process.
2. We propose the variable bypass bin incorporation (VBBI) technique, which allows an extra two bypass coding bins to be processed in the same clock cycle in a quad parallel architecture, significantly improving the throughput of BAE in a parallel architecture.
3. When the context model cannot be prefetched early enough, the pipeline will stall since the context model needed for the current bin typically depends on the outcome of the previous bin. We provide a prediction-based context model prefetching strategy to address this issue. Additionally, the Multi-result Context Model Update (MCMU) architecture is proposed, the critical path for state transitions is shortened by the context model update of the meticulously optimized parallel architecture.
4. Based on the HEVC video coding standard, a highly compatible hardware architecture for entropy encoding is provided. The whole entropy encoding architecture is pipelined, and the data interaction between binarization and BAE is cached using parallel-in-parallel-out (PIPO) to improve the stability of the entropy encoder. It also develops a quad-loop cache architecture to improve compatibility for data interaction between the entropy encoder and other video encoder modules.

2. Analysis of CABAC

2.1. CABAC's Process

As depicted in Figure 1, CABAC comprises three key modules: binarization, context modeling, and binary arithmetic coding [23]. The video prediction data, reference data, etc., are parsed into the appropriate SEs in the entropy coding process. These SEs include prediction patterns, block segmentation flag, etc. After binarization, the values of the non-binarized SEs are mapped into a series of a variable number of binary symbols [22]. Each binary symbol is referred to as a bin. The critical information of the video sequence is represented by the syntax elements, which aim to represent the video with the least amount of data possible while allowing for the reconstruction of the video sequence at the decoding stage.

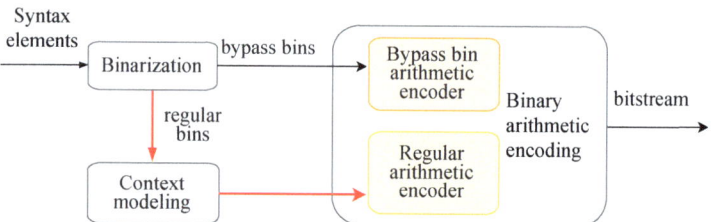

Figure 1. Key components of CABAC.

The binary symbol bin is the data that can be processed directly by the arithmetic coding module. Arithmetic coding is primarily split into Regular Coding and Bypass Coding, with various SEs accessing distinct selection criteria for each. Among them, the context modeling part will supply the context probability model of the associated bin based on the context data from the SEs for the regular coding bin.

The HEVC standard defines several binarization methods for entropy coding: Fix-Length (FL) coding, Truncated Rice (TR) coding, K-order exponential Golomb coding, etc. The above binarization methods are the most critical for syntax elements in HEVC, except for very few syntax elements with their own specific binarization methods. This is mainly influenced by the numerical characteristics of different SE values and is related to the context model selection methods corresponding to other SEs. In addition, although the binarization method of SEs is specified directly by the standard, the quantization parameters cMax and RiceParam often depend on the specific encoding situation. For example, the cMax parameter of the merge_idx index is determined by the number of merge mode candidates.

The probability of encoded blocks and encoded SEs is reflected in the context model in entropy coding. The core of context modeling is to estimate the distribution probability of the currently encoded SEs and enhance coding efficiency by using video data's spatial and temporal correlation. The accuracy of context modeling, which holds a key place in the entropy coding standard, significantly affects the coding effect. For the standard coding model, the coding procedure for each bin includes the corresponding context model. To adaptively make adjustments to diverse videos, these context model need to be updated in real time.

Although the arithmetic coding specified by the HEVC standard is conceptually comparable to the joint finite-precision binary arithmetic coding, numerous modifications have been made to the implementation techniques to reduce the complexity of the operations. The More Probable Symbol (MPS) and Less Probable Symbol (LPS) definitions of the encoding's binary symbols denote the symbols having a big and small probability of occurrence, respectively. The binary arithmetic encoding inputs are the bin to be encoded and its accompanying contextual model. Figure 2 depicts the encoding procedure, primarily separated into the MPS and LPS bin types. Although the two flow lines are different, they include stages like renormalization, calculating rLPS and updating the context.

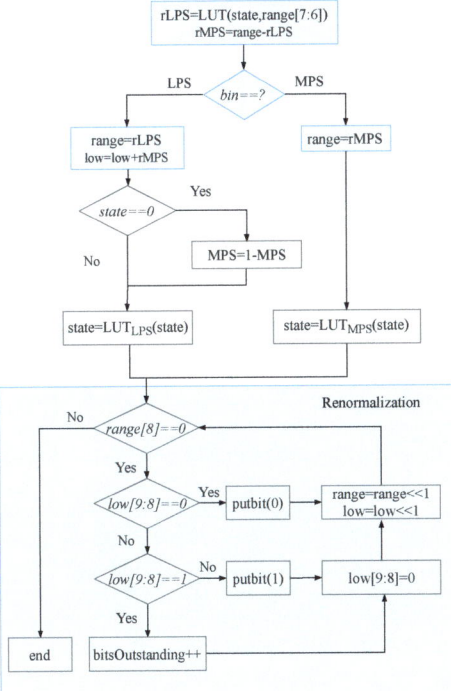

Figure 2. Flowchart of BAE. The gray section can be pre-executed before the blue section.

2.2. Bottleneck Analysis

The pipeline architecture is one successful approach to increasing the throughput of BAE hardware, and the multi-channel parallel architecture is another. BAE in HEVC suffers from a huge area of memory due to lots of context models [24,25], so Static Randomaccess Memory (SRAM) is used instead of registers. However, a particular case in the implementation causes the pipeline architecture to stall. As shown in Figure 3, when the current bin coding is complete, the context model of the same bin must be restored for the next bin at the next clock cycle. Updating the context model requires one clock, and reading or writing the context model from RAM also consumes one clock, so subsequent bins cannot read the updated context model from the adjacent clock from the context model RAM that has not yet been written. Therefore, it is necessary to suggest a CABAC hardware design that can implement a parallel or pipelined CABAC without stalling.

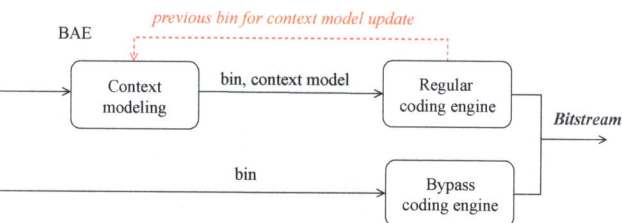

Figure 3. Each time a bin is encoded in the regular coding engine, the context model must be changed and saved back into the Context Modeling.

For multiplexed parallel context model update architectures, the resulting path delay corresponds to many levels of multiplexers, which will dominate the critical path of CABAC. Thus, bottlenecks have emerged in determining how to improve the efficiency of the pipeline/parallel structure, and use less hardware to achieve better throughput CABAC designs.

For a variety of data, other video encoder modules communicate with the entropy encoding. Numerous data will be combined in the entropy coding. The entropy coding may occasionally fail to finish digesting the input data in a timely manner, resulting in the loss of the input data since its coding efficiency differs from that of the other modules of the encoder. The residual coefficient data are the largest class of data among the coded data required for entropy coding. It also becomes challenging to balance the data supply of the reconstruction module with the value of the entropy coding and how to store these data more effectively.

3. Proposed CABAC Prediction-Based Context Model Prefetching Strategy

3.1. Prediction-Based Context Model Prefetching

One of the features of CABAC is that each time a regular encoding is performed, the probabilistic model of the current encoded bin needs to be updated. The context modeling needs to transmit the same throughput to support the BAE with multiple bins constructed above.

However, pipeline or parallel implementation is complex when faced with some exceptional cases. When there are successive bins with the same context model, since one clock is required to update the context model and both reading and writing of RAM data also occupy one clock, the latter bin cannot be read from the context model memory CM_RAM in the adjacent clock cycle that has not yet been written to the updated context model. To cope with the phenomenon of pipeline stall, this paper proposes a context model prefetching strategy and optimizes it for the multi-bin case, aiming to achieve a stall-free pipeline and low resource and high master frequency.

The context modeling architecture of the pipeline BAE in this paper is shown in Figure 4a. Because only one bin is processed per cycle, the design of this paper uses Parallel In Serial Out (PISO) as the input module for context model update. The PISO module outputs data for one bin at each clock cycle. At the same time, the context model needs to be obtained from CM_RAM by index. Prefetching will save the relevant data and predict the next incoming bin to be the same context model as the current bin. Finally, the predicted bin values and other data are transferred to the next stage. Since RAM reading and writing consume one clock cycle, if consecutive bins utilize the same context model, the post-context model cannot access the data written after the pre-update of the adjacent clock cycle.

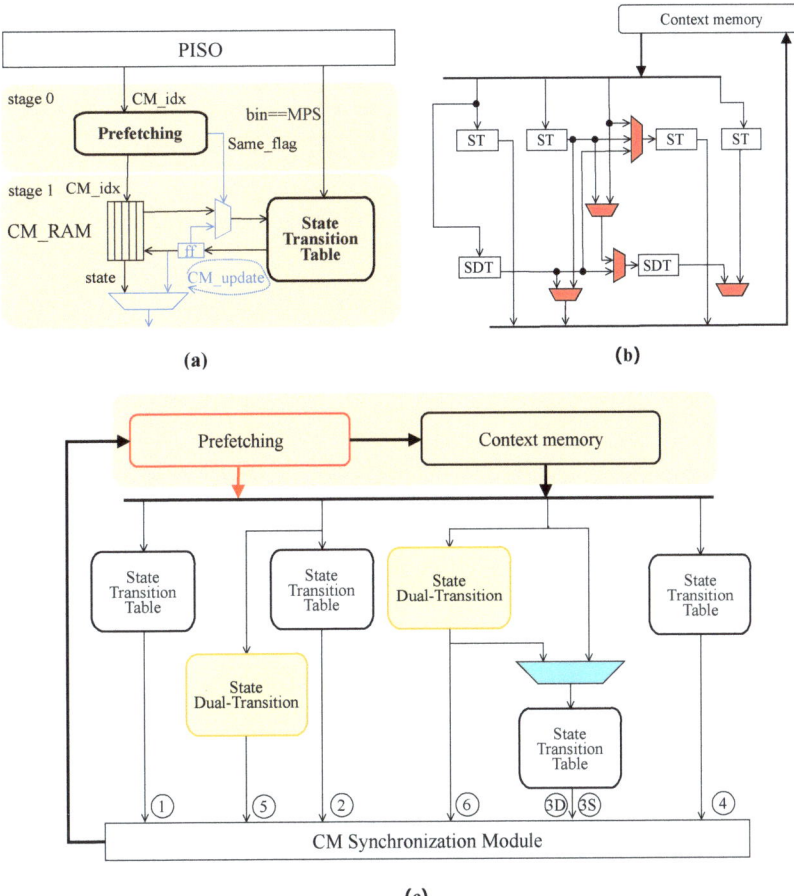

Figure 4. Proposed SDT-based context updating architecture. ST and SDT are 128-1 and 256-1 LUTs, respectively. (**a**) Proposed architecture for updating single-way pipeline context model based on prefetching. (**b**) Architecture proposed by [18] with additional state dual-transition (SDT) LUTs. (**c**) Proposed MCMU architecture. The values from 1 to 6 denote the possible outputs.

When the index of the current clock cycle input to CM_RAM is the same as the previous clock cycle, the context model is directly communicated to stage 1 of the pipeline through the Same_flag, and the context model is directly passed through the internal pipeline. The CM update module receives the updated model in the previous clock cycle instead of using CM RAM. Thus, regardless of whether the context model of the next bin is the same as the current consistent one, the correct data can be output promptly. The context model that needs to be updated is found in the state transition table and then saved in CM_RAM for real-time updates. This is a prefetching strategy proposed in this paper to solve this case, implemented by caching the model's index.

As depicted in Figure 4b, ref. [18] designed an architecture for context model update in parallel architecture. However, the critical path must be further optimized to prevent the critical path delay from exceeding BAE. To address this issue, we propose the Multi-result Context Model Update (MCMU) architecture in this paper.

Ref. [18] had to make the probabilistic state update satisfy all bin cases; many multiplexers are placed between state transition (ST) and state dual-transition (SDT), which is considered to be simplified in this paper, and the new architecture is shown in Figure 4c. The architecture utilized in this work features one clock cycle for encoding up to four bins, where the type of bin specified in Table 1 denotes the interrelationship of the context models of these four bins. When the bin type is the same, as indicated by the same context model, it yields a total of only seven cases. For instance, if the bin type is ABBD, this means that the middle two bins utilize the same context model. The proposed architecture provides six results per clock cycle. Among them, result 3 contains two cases that must be arbitrated by prefetching. Therefore, this architecture can obtain all the results of the context model update by only one multiplexer, at the cost of dropping the encoding of the last bin if all the four bins are of the same context model, i.e., only three bins are encoded in parallel in this clock cycle.

Table 1. All cases with the context model dependencies for the four-way parallel bin.

Bin type	A	B	C	D
Forms used	ST	ST	ST	ST
The result of use	1	2	3S	4
Bin type	A	A	C	D
Forms used	ST	SDT1	ST	ST
The result of use	1	5	3S	4
Bin type	A	A	A	D
Forms used	ST	SDT1	SDT2 + ST	ST
The result of use	1	5	3D	4
Bin type	A	A	A	A
Forms used	ST	SDT1	SDT2 + ST	QT
The result of use	1	5	3D	NG
Bin type	A	B	B	D
Forms used	ST	ST	SDT2	ST
The result of use	1	2	6	4
Bin type	A	B	B	B
Forms used	ST	ST	SDT1	SDT2 + ST
The result of use	1	2	5	3D
Bin type	A	B	C	C
Forms used	ST	ST	ST	SDT2
The result of use	1	2	3	6

Suppose there are bins with the same context model in the next clock cycle. In that case, the context model needs to transfer to the BAE module first through the cache in time to avoid the untimely transmission of the context model due to the read and write time of the memory.

3.2. Proposed Pre-Range Update and Pre-Renormalize BAE Architecture

The context model update, computation of *range*, calculation of *low*, and the renormalization procedure are all carried out in a cascading manner, as is already noted, making the entire coding process feasible for pipelining activities.

A valuable fact for designing a single-way arithmetic coding pipeline is that the *low* of the current coding interval depends on *range* unidirectionally. In contrast, *range* does not depend on *low*. Therefore, in the design of the arithmetic coding pipeline, *range* and *low* can be calculated separately, and *range* is calculated first before *low* to shorten the critical path of the pipeline.

According to Figure 2, the *range* computation includes rLPS lookup, interval recursion, renormalization lookup and shift; *range* update depends on rLPS, and rLPS depends on the current *range* and context model state. Therefore, compared with the *low* update and bitstream output part, the computation of *range* is the most complex part of the pipeline, and the loop algorithm generated by renormalization becomes a bottleneck for the hardware architecture.

However, one renormalization can only double the encoding interval of less than 256, so it is often necessary to perform multiple renormalizations due to the small *range*, which makes it difficult to pipeline and affects the encoding efficiency. Since only the shifting of *range* and *low* and the counting of *bitOutstanding* are performed in the renormalization process, multiple renormalizations can be completed in one operation. The times of renormalizations differ when the bin is MPS and LPS. When the bin is MPS, if *range* is less than 256, then renormalization is performed once; otherwise, renormalization is skipped. Renormalization is required when the bin is LPS, as illustrated in Table 2; renorm count is retrieved from the renormTab table. The lookup table uses the higher five bits of the rLPS as an index for the times of renormalizations.

Table 2. Renormalization times table.

rLPS [7:3]	0	1	2–3	4–7	8–15	16–31
Renormalization times	6	5	4	3	2	1

Therefore, this paper proposes pre-range update and pre-renormalize, as shown in the purple area of Figure 5. In the first stage, in addition to completing the pre-computation rLPS, the pre-lookup table and storage structure of the renormalization count renorm_count are added. The *range* update is split into two levels of pipeline. Renorm_count is obtained from rLPS by indexing the table. The renormTab table size is 1×32, so in the context of the first stage pipeline to obtain four candidate rLPSs, we can also look up the table to obtain the renormalization number renorm_count which corresponds to the candidate rLPS, and shift to obtain the corresponding renormalization interval rLPS_renorm with four candidate values. So the renormalization count lookup table of rLPS can also be split into sub-operations carried out in the first stage pipeline. In contrast, the shift operation in renormalization is completely placed in the first stage pipeline.

The four candidate renormalization values obtained after pre-renormalize will also be used as indexes by *range* [7:6] at the beginning of the second stage pipeline to determine the final rLPS renormalization interval. For the renormalization of MPS, the above pre-normalization method cannot be used because it depends entirely on the coding interval *range* of the previous encoding. In the second stage of the pipeline, the highest bit of rMPS is used as the judgment condition to determine whether to perform rMPS renormalization. The final *range* is selected between rLPS_renorm and rMPS_renorm according to whether the bin is MPS or not.

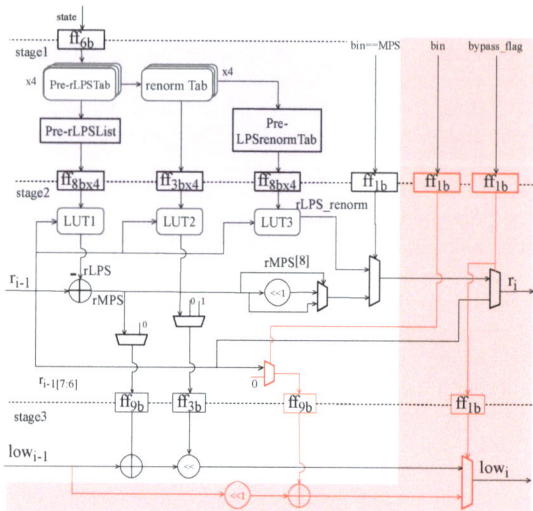

Figure 5. Diagram of pipelined single-bin BAE architecture. The purple part is the proposed pre-renormalize technique. The red part is the single-way hardware architecture which is bypass bin compatible.

3.3. Area-Efficient BAE Pipeline Architecture with Compatibility

In the entropy coding of HEVC, the bins of regular and bypass coding are sequentially arranged. If hardware is designed separately for both, it can achieve very high coding speed under certain circumstances, especially for bypass bins. The bypass bin splitting (BPBS) described by [18], which increases the throughput of 1 clock cycle, has several implementation limitations. Still, at the cost of memory, resources to store intermediate results and additional bin sequences merge to integrate. For up to five pathways of [18] alone, 32 combinations of bin cases are included. It is also required to allocate all the results in one clock cycle before the update of *low* can be performed. If more multiple bypass bins are attempted, more cases need to be processed, which will be an extremely complex process that will consume a large hardware area and may become a new critical path. So this paper proposes area-efficient BAE pipeline architecture with compatibility.

The coding state is calculated differently for different coding modes, as shown in Table 3. The bypass flag of the current bin is stored in the second stage of the pipeline to select the encoding state computed in different encoding modes. The bypass encoding *range* remains unchanged, with only *low* changes. Our work integrates the bypass and regular encoding in one hardware architecture. The update of *range* is currently the critical path, so combining the update of bypass coding *low* into the pipeline does not cause the frequency to decrease.

Table 3. Low update with different coding methods.

Coding Method	Input	Low Update
Regular	MPS	$low \ll renorm\_count$
Regular	LPS	$(low + rMPS) \ll renorm\_count$
Bypass	1	$low \ll 1 + range$
Bypass	0	$low \ll 1$

The third stage of the pipeline architecture designed in this paper is the update calculation of *low*. The number of renormalizations when the bin is LPS is obtained via renormTab and LUT2 jointly checking the table. If the bin is MPS and rMPS is greater than or equal to 256, the renormalization is skipped, and if the bin is MPS but rMPS is less than

256, the renormalization is carried out once. The bypass coding only updates *low*, and its coding process is shown in the red part in Figure 5. Compared to the design presented in the previous work, the current architecture can accomplish stable and continuous coding with a lower circuit area without extending the critical path.

3.4. Multi-Bin Parallel Architecture Based on Variable Bypass Bin Incorporation

A pipeline or parallel architecture are two efficient ways to increase the throughput rate of arithmetic coding technology. However, the dependency on the arithmetic encoding states makes the issue of long-timing routes in the pipeline structure even worse. This work presents a pipeline architecture for the arithmetic encoder and a multi-path parallel architecture with a single pipelined arithmetic encoder on each lane.

In the four-way parallel structure shown in Figure 6, the context model updates are precomputed upfront. The *range* and *low* computed by the first encoder are used as the state input for the second channel, and so on for multiple channels of state updates. In particular, the encoding state of the last encoder will be saved in a register as the starting state data for the next set of four-way bin encoding.

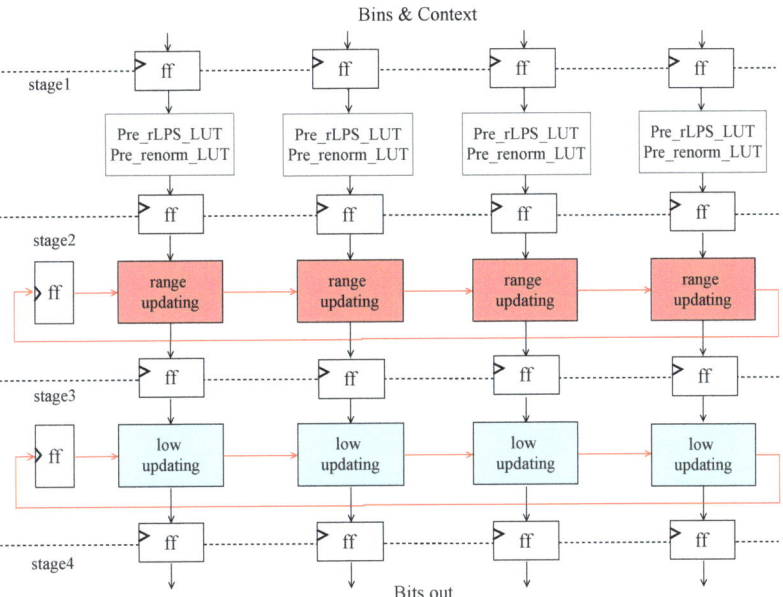

Figure 6. Pipelined multibin BAE architecture.

In the first stage of the pipeline, in addition to the pre-rLPScalculation and pre-renormalize, a pre-lookup table and storage structure for the renormalization count are added. The renormalization count candidates are stored in registers and will determine the final value in the second-stage pipeline, which will participate in the renormalization calculation of *low*. In a basic four-bin BAE, either a regular or a bypass bin must be encoded sequentially. As we can see through the previous section, the update phase of *range* in a single-path pipeline architecture becomes the critical path, while the update of *low* is more straightforward.

So this paper proposes the Variable Bypass Bin Incorporation (VBBI) architecture, as shown in Figure 7. By taking advantage of the feature that the bypass bin does not change the context model and *range*, each time four bins are encoded, if immediately followed by one or two bypass bins, these two bins are added to the current bin sequence to achieve the maximum throughput rate of six bins encoded in at most one clock cycle. Even if the update of *low* increases to six bins at the same time, the critical path does not exceed the update process of the *range* of four bins, so there is no impairment in the main frequency performance, and the throughput of the parallel architecture can be effectively increased. Compared with the bypass bin separation architecture used in [18], every single path in this paper can be adapted to bypass coded bins, which not only saves the RAM used to store intermediate variables but also removes the hardware area generated by using bin sequence merge, and the average throughput rate can be achieved very close.

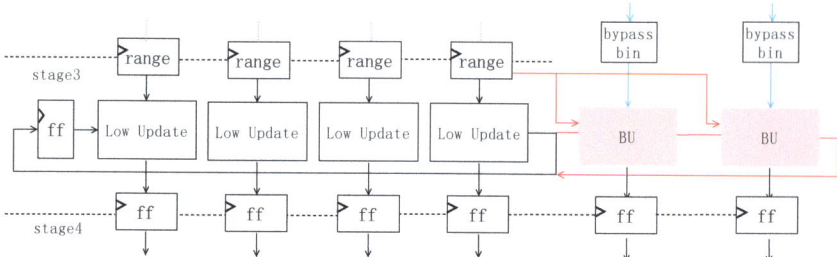

Figure 7. Variable Bypass Bin Integration (VBBI) architecture. It can encode up to 6 bins in 1 clock, including at least two bypass bins.

4. Overall System

4.1. Quad-Loop Cache Input

Entropy coding of HEVC is a module that performs statistically based lossless data compression of the results generated by other modules, so it is related to each module in video coding. The coding framework of the entropy coding module is shown in Figure 8. When the entropy coding module obtains all the SEs and residual coefficients, it needs to pre-process the syntax elements and residual coefficients at each level, which includes calculating the values of syntax elements to be coded, the context model index, and the coding method. After the SEs are generated, they will enter the binarization core and input the binarized bin into the PIPO memory. Then the prefetching module will input three to six bins per clock cycle into the BAE, which the bit generator will finally integrate into the bitstream output.

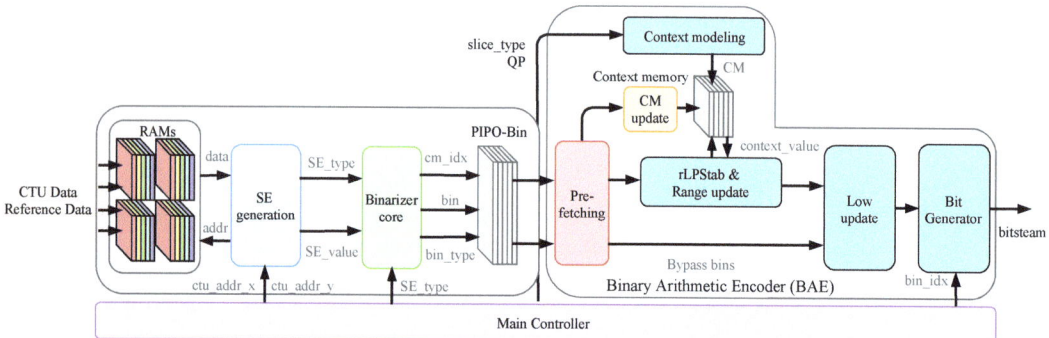

Figure 8. Overall framework of a highly compatible VLSI architecture for H.265/HEVC CABAC encoder for UHD TV applications.

The data in the entire CTU generated in the video encoding process are passed to the entropy coding module; these data are diverse and need to be considered for hardware architecture to match the timing of transmission. However, the data processing speed of other modules and this module's throughput will differ. Under the condition that the whole video coding is pipelined architecture, the data input structure, as shown in Figure 9, is used to enhance the compatibility of entropy coding. The quad-loop cache architecture is different from the First-Input-First-Output (FIFO) memory in that it completes the FIFO function for each group of RAMs, and the data in the RAM block can be read out in disorder, which is suitable for the data reading requirement of the entropy coding module. If Drw is 4, Write Pointer (WP) is one turn ahead of Read Pointer (RP) and points to the same RAM as RP. If the pipeline continues to run, it will lead to data loss and coding errors. Therefore, when Drw is equal to 4, the rest of the video encoding process needs to be paused to ensure that the coding is absolutely correct.

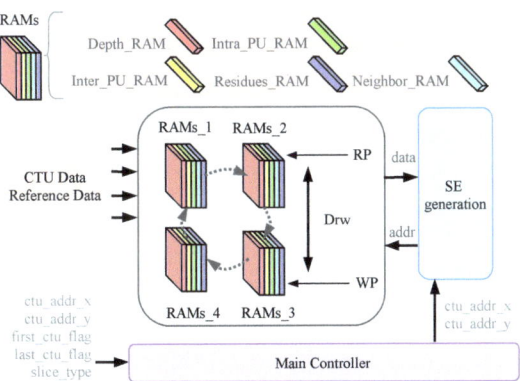

Figure 9. Quad-loop cache architecture.

The input data include Depth_RAM (containing information on CU depth, TU depth, and PU mode), Intra_PU_RAM (luminance and chrominance direction), Inter_PU_RAM (information related to merge and amvp), Neighbor_RAM (information related to the top side and left side CTU), Residues_RAM (residual data), etc. The data to be entropy coded are cyclically cached through four RAMs. Each group of RAMs keeps all the data of one CTU, effectively reducing the dependency between video coding modules.

4.2. Binarization Architecture

The binarization schemes used for most of the SEs in HEVC are Truncated Unary (TU), Truncated Rice (TR), Kth-order ExpGolomb (EGK), and Fixed-Length (FL) codes. The rest of the SEs use their corresponding custom binarization schemes, which will include some compound encoding [26].

Since the binarization is carried out separately for each SE and is not the bottleneck of the whole architecture, as long as the average throughput of the part is higher than the average throughput of the BAE, in any case, the entire architecture can be satisfied with smooth and efficient operation.

The architecture of the single-core binarization module is shown in Figure 10. The input is SE encoding type value, which is encoded according to the respective encoding rules. The output of the completed encoding are the bin value, the context model index, and the encoding type [27].

This design uses a parallel three-stage binarization scheme to meet the goal of smooth and efficient binarization, as shown in Figure 11. The first stage is responsible for inputting and sorting the syntax element values SE_Value and encoding types SE_Type that need to be binarized in order and then transferring them to the following encoding stage [28]. The second stage is responsible for binary encoding. It consists of two single-core binary modules, one combined module, and one custom module. Each single-core binary module supports four binary schemes, and the four modules are independent of each other. The third stage is to type each data after binarization into a packet containing the current bin value, the coding type, and the contextual model index. These data are then integrated into the PIPO module and passed into the arithmetic encoding and the context model module as required to achieve a pipeline architecture for the entire entropy encoding module [22].

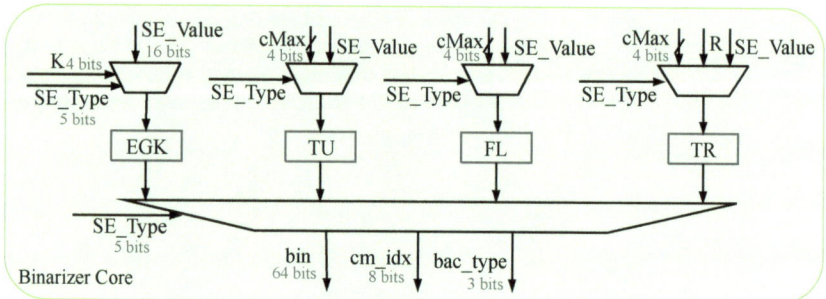

Figure 10. Proposed single-core binarization architecture.

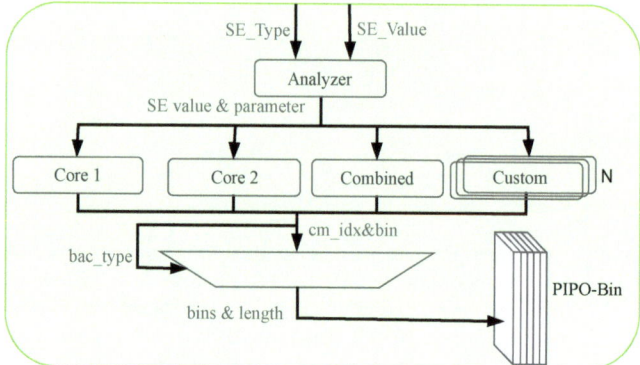

Figure 11. Proposed parallel binarization architecture. It can implement binarization of multiple syntax elements in 1 clock.

5. Implementation Results

Experiments are conducted to evaluate the performance of the proposed architecture, and the superiority of the proposed CABAC encoder is tested via the HEVC reference software HM-16.7. The proposed CABAC encoder is implemented in Verilog HDL. RTL simulation is performed on 18 sequences in 5 classes. Tests cover All Intra (AI), Low Delay (LD), Low Delay P, and Random Access (RA) configurations and include settings for Quantization Parameters (QPs) 22 and 37.

The CABAC pipeline 1 bin/clock architecture designed in this paper avoids the pipeline stall problem. Table 4 presents the encoding time that can be saved when encoding a video sequence since the approach in this paper avoids the stall of the pipeline architecture caused by successive identical context models. Under general test settings in the AI configuration, the suggested CABAC architecture can save up to 45.66% of the coding time by employing the prediction-based context model prefetching method. Even in the LD, LD_P, and RA settings, the encoding time can have significant reductions. When the QP is low, the encoding time can be reduced by 27.5% on average, and even when the QP is 37, the pipeline architecture stalls can be optimized by 20.95% on average. This is because the context model prefetching architecture proposed in this paper can be adapted to the pipeline architecture to avoid the time consumption caused by the context model update in memory. The time savings differ since low QP values for high-resolution video increase the SEs associated with coding residuals. These SEs provide many bins with the same contextual model for standard coding.

Table 4. Percentage of encoding time (%) saved by the CABAC pipeline architecture improving the proposed prediction-based contextual model prefetching strategy under common test conditions.

Class	Sequence	All Intra (AI)		Low Delay (LD)		Random Access (RA)		Average	
		qp = 22	qp = 37	qp = 22	qp = 37	qp = 22	qp = 37	qp = 22	qp = 37
A	PeopleOnStreet	35.50	25.56	30.78	18.47	29.03	18.96	31.28	20.28
	Traffic	30.80	29.76	22.73	20.63	23.66	23.02	24.75	23.30
B	ParkScene	34.86	32.70	25.68	21.23	26.58	23.27	28.01	24.29
	Kimono1	**45.66**	37.45	38.03	28.25	38.29	28.82	39.71	30.39
	BasketballDrive	34.54	25.84	33.05	23.94	34.05	23.94	33.08	23.42
	BQ Terrace	38.38	27.96	32.63	23.29	32.99	24.49	33.97	24.33
	Cactus	31.47	27.70	30.34	21.82	28.14	22.74	29.64	23.38
C	BasketballDrill	22.01	21.19	23.54	17.90	22.03	18.07	22.57	18.70
	BQ Mall	25.16	25.30	23.19	18.24	23.19	19.66	23.55	20.31
	PartyScene	23.30	22.29	21.43	18.40	22.10	18.18	21.95	19.13
	Race Horses	35.38	27.48	29.24	17.95	28.35	19.81	30.56	20.81
D	BasketballPass	26.16	22.20	22.27	17.75	22.00	18.04	23.10	18.92
	Blowing Bubbles	22.80	22.98	19.44	16.71	21.30	18.23	20.68	18.67
	BQSquare	27.08	21.58	18.44	11.96	17.85	15.33	20.36	15.16
	Race Horses	32.46	21.59	21.94	16.72	22.92	17.04	24.81	18.02
E	Kristen And Sara	29.38	24.30	29.93	14.84	28.88	18.63	29.48	18.07
	FourPeople	27.39	26.22	26.85	16.39	26.73	20.78	26.81	19.91
	Johnny	33.03	26.30	30.08	17.08	30.57	20.05	30.76	20.08
	Average	**30.85**	26.02	26.64	18.98	26.59	20.50	**27.50**	20.95

This paper's architecture follows the anticipated strategy, allowing it to avoid pipeline standstill brought on by context model updates and allocate the number of codes per group of bins through the prefetching module in the parallel architecture, which significantly increases coding efficiency. Our proposed CABAC encoder is implemented in Verilog. RTL simulations are performed using the bin sequences in Table 4 across five different resolution classes, and the QPs are 22 and 37. Table 5 shows the effect of LCMU in the simulation. With the LCMU, the number of delivered bins per clock cycle (BPCC) is slightly reduced to below 4, but the maximum clock frequency is substantially increased. Further using VBBI, the final BPCC is between 4.10 and 4.39 (depending on the configuration). Table 6 shows the probability that the parallel architecture stalls due to untimely model reads caused by the next set of bins having the same contextual model as the previous set, the coding time that can be optimized via the prefetching architecture proposed in this paper.

For the full pipelined architecture CABAC, the gate count is 39.52 K, the maximum operating frequency is 714 MHz, and the maximum throughput is 714 Mbin/s. For the CABAC with the highly compatible parallel architecture in this paper, the overall CABAC throughput, at 513 MHZ, is 2191 Mbin/s. Numerous predictive lookup tables and alternative algorithms are required to raise the frequency and the number of parallel bins, and these efforts have led to higher throughput. Therefore, the throughput rate is also the highest due to the optimization of the hardware design and more advanced processes in this work.

Table 5. Percentage of coding time (%) saved via the prediction-based context model prefetching strategy proposed by the CABAC parallel architecture improvement.

Sequence	Config.	qp = 22	qp = 27	qp = 32	qp = 37
BasketballDrive	LD	30.39	25.51	22.27	18.52
	RA	30.04	24.46	21.30	18.07
Traffic	LD	22.09	22.36	20.96	18.83
	RA	22.31	23.04	22.27	20.63
PeopleOnStreet	LD	27.84	22.56	18.69	15.86
	RA	25.54	21.19	18.08	15.80
BQTerrace	LD	33.45	27.03	25.49	22.31
	RA	32.80	26.34	23.83	22.89
Kimono	LD	34.55	32.20	29.62	26.33
	RA	33.52	31.12	28.46	25.61
Average	24.60 *	29.25	25.58	23.10	20.48

* Overall average.

Table 6. Performance in number of delivered BPCC for H.265/HEVC.

Sequence	Config.	LCMU	LCMU + VBBI
BasketballDrive	LD qp = 22	3.90	4.23
	LD qp = 37	3.96	4.28
	RA qp = 22	3.90	4.26
	RA qp = 37	3.96	4.33
Traffic	LD qp = 22	3.93	4.22
	LD qp = 37	3.96	4.25
	RA qp = 22	3.93	4.29
	RA qp = 37	3.94	4.29
PeopleOnStreet	LD qp = 22	3.90	4.32
	LD qp = 37	3.97	4.37
	RA qp = 22	3.91	4.38
	RA qp = 37	3.97	4.42
BQTerrace	LD qp = 22	3.86	4.10
	LD qp = 37	3.93	4.23
	RA qp = 22	3.87	4.15
	RA qp = 37	3.93	4.27
Kimono	LD qp = 22	3.86	4.26
	LD qp = 37	3.91	4.21
	RA qp = 22	3.87	4.34
	RA qp = 37	3.91	4.28
Average		3.92	4.27

Table 7 summarizes the design specifications of our CABAC encoder, compared with the state of the art. Many authors have reported on the CABAC architecture in the past, with their focus on different ASIC technologies. The pipeline architectures of [10–12] have similar configurations. Nonetheless, this paper achieves higher throughput rates and smaller circuit areas by targeting critical path optimization for the renormalization part of the *range* update, context model accesses using RAM only, binarization using a single core and more advanced ASIC processes. Ref. [16] parallel architecture designs use the bypass bin splitting technique and merge bypass bins, respectively, to increase the throughput rate. Although their bins per clock cycle are slightly higher than this paper, they pay a high price, such as adding bins splitting/merging modules and PIPOs for storing data such as ranges, etc. The proposed architecture in this research enhances the frequency using the pre-renormalize technique and MCMU, while increasing the throughput by utilizing VBBI, ultimately resulting in improved hardware efficiency, as illustrated in Figure 12. Specifically, the hardware efficiency (Mbins/s per k gates) achieved in this paper is higher than that of other architectures, including both pipeline and parallel architectures; this work achieves 20.16 Mbins/s per k gates. Additionally, the context model prefetching strategy employed in this paper effectively eliminates the time delay that arises due to model updates in memory, effectively mitigating the BAE stagnation problem.

Table 7. Specification and comparison with prior arts.

Design	Kim [12]	Peng [13]	Ding [14]	Zhou [18]	Zhang [29]	This Work	
Process/nm	IDEC 180	TSMC 130	Kintex-7	TSMC 90	TSMC 90	TSMC 65	
gate count/K	45.089	48.94	-	110.9	54.5	39.52	108.7
Max.clock frequency/MHZ	158	357	120	420	720	714	513
bins/clock	1	1.18	3.59	3.29 (4.37) *	-	1	4.27
Max·throughput/Mbin·s^{-1}	158	261–439	431	1382 (1836) *	850	714	**2191**
Mbin·s^{-1}/gate count	3.5	5.33–8.97	-	12.48 (16.56) *	15.6	**18.07**	20.16

* The actual results, the ones in parentheses are the occasional optimal results.

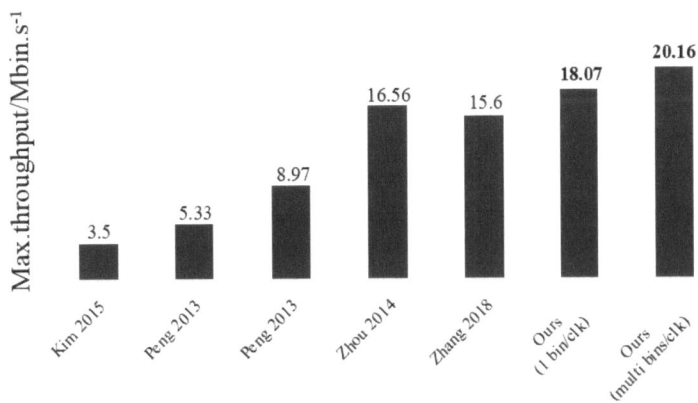

Figure 12. The proposed optimizations improved the hardware efficiency of CABAC when compared to existing papers [12,13,18,29].

6. Summary

The occurrence of consecutive bins in the same context model can cause stalls in the hardware pipeline architecture. To overcome this problem, we propose a prediction-based context model prefetching strategy to alleviate data dependencies by predicting the next bin model, and reduce critical path delays through the MCMU. In addition, we use pre-range update and pre-renormalize technique to reduce the multiplex BAE's route delay due to the incomplete reliance of the encoding process. Then, we propose the VBBI technique to improve the throughput of BAE in a parallel architecture. Moreover, the data interaction between CABAC modules is optimized. In accordance with the experiments, our architecture eliminates pipeline stalls and saves encoding time, and works better for high resolution and low QP values, which is in line with the need for more high-definition videos as time progresses. Moreover, the throughput is enhanced and the hardware efficiency of the pipeline architecture is maximized. In future study, we will focus on making this work compatible with the multi-channel parallel architecture and Versatile Video Coding (VVC/H.266) hardware design [30].

Author Contributions: Conceptualization, methodology, software, writing—review and editing, C.F.; data curation, experiments, H.S.; investigation, experiments, data curation, writing—review and editing, Z.Z.; supervision, J.Z. All authors have read and agreed to the published version of the manuscript.

Funding: This work was supported by JSPS Bilateral Programs Joint Research Projects Grant Number JPJSBP120 223210, and this work was supported through the activities of VDEC, The University of Tokyo, in collaboration with NIHON SYNOPSYS G.K, and Japan Society for the Promotion of Science (JSPS), under Grant 21K17770.

Institutional Review Board Statement: Not applicable.

Informed Consent Statement: Not applicable.

Data Availability Statement: Data sharing is not applicable to this article.

Conflicts of Interest: The authors declare no conflict of interest.

Abbreviations

The following abbreviations are used in this manuscript:

IoT	Internet of Things
CABAC	Context Adaptive Binary Arithmetic Coding
HEVC	High Efficiency Video Coding
MCMU	Multi-Result Context Model Update
VBBI	Variable Bypass Bin Incorporation
QP	Quantization Parameter
UHDTV	Ultra High Definition Television
JCT-VC	The Joint Collaborative Team on Video Coding
CTUs	Coding Tree Units
CAVLC	Context-based Adaptive Variable Length Coding
RDO	Rate Distortion Optimization
AVC	Advanced Video Coding
SEs	Syntax Elements
BEE	Bypass Encoding Engines
MBBS	Multiple Bypass Bins Scheme
BAE	Binary Arithmetic Encoding
RTL	Register Transfer Level
MRSET	Multiple Residual Syntax Element Treatment
PIPO	Parallel-In-Parallel-Out
FL	Fix Length
TU	Truncated Unary
TR	Truncated Rice
MPS	More Probable Symbol
LPS	Less Probable Symbol
SRAM	Static Randomaccess Memory
SDT	State Dual-Transition
PISO	Parallel In Serial Out
ST	State Transition
BPBS	Bypass Bin Splitting
FIFO	First-Input-First-Output
WP	Write Pointer
RP	Read Pointer
AI	All Intra
LD	Low Delay
RA	Random Access
BPCC	Bins Per Clock Cycle
VVC	Versatile Video Coding

References

1. Bhering, F.; Passos, D.; Ochi, L.S.; Obraczka, K.; Albuquerque, C. Wireless multipath video transmission: When IoT video applications meet networking—A survey. *Multimed. Syst.* **2022**, *28*, 831–850. [CrossRef]
2. Jiang, X.; Yu, F.R.; Song, T.; Ma, Z.; Song, Y.; Zhu, D. Blockchain-enabled cross-domain object detection for autonomous driving: A model sharing approach. *IEEE Internet Things J.* **2020**, *7*, 3681–3692. [CrossRef]
3. Cisco, V. Cisco visual networking index: Forecast and trends, 2017–2022. *White Pap.* **2018**, *1*, 1–30.

4. Sullivan, G.J.; Ohm, J.R.; Han, W.J.; Wiegand, T. Overview of the high efficiency video coding (HEVC) standard. *IEEE Trans. Circuits Syst. Video Technol.* **2012**, *22*, 1649–1668. [CrossRef]
5. Flynn, D.; Marpe, D.; Naccari, M.; Nguyen, T.; Rosewarne, C.; Sharman, K.; Sole, J.; Xu, J. Overview of the range extensions for the HEVC standard: Tools, profiles, and performance. *IEEE Trans. Circuits Syst. Video Technol.* **2015**, *26*, 4–19. [CrossRef]
6. Zhang, J.; Kwong, S.; Zhao, T.; Pan, Z. CTU-level complexity control for high efficiency video coding. *IEEE Trans. Multimed.* **2017**, *20*, 29–44. [CrossRef]
7. Adireddy, R.; Palanisamy, N.K. Effective approach to reduce complexity for HEVC intra prediction in inter frames. In Proceedings of the 2014 Twentieth National Conference on Communications (NCC), Kanpur, India, 28 February 2014–2 March 2014; pp. 1–5.
8. Huang, B.; Chen, Z.; Su, K.; Chen, J.; Ling, N. Low-Complexity Rate-Distortion Optimization for HEVC Encoders. *IEEE Trans. Broadcast.* **2021**, *67*, 721–735. [CrossRef]
9. Choi, J.A.; Ho, Y.S. Improved residual data coding for high efficiency video coding lossless extension. In Proceedings of the the 2013 RIVF International Conference on Computing & Communication Technologies-Research, Innovation, and Vision for Future (RIVF), Hanoi, Vietnam, 10–13 November 2013; pp. 18–21.
10. Wang, M.; Bi, Q.; Zhu, Y. Video compression: A jointly optimized transform-quantization method. In Proceedings of the 2017 IEEE International Conference on Signal Processing, Communications and Computing (ICSPCC), Xiamen, China, 22–25 October 2017; pp. 1–5.
11. Wiegand, T.; Sullivan, G.J.; Bjontegaard, G.; Luthra, A. Overview of the H. 264/AVC video coding standard. *IEEE Trans. Circuits Syst. Video Technol.* **2003**, *13*, 560–576. [CrossRef]
12. Kim, D.; Moon, J.; Lee, S. Hardware implementation of HEVC CABAC encoder. In Proceedings of the 2015 International SoC Design Conference (ISOCC), Gyeongju, Republic of Korea, 2–5 November 2015; pp. 183–184.
13. Peng, B.; Ding, D.; Zhu, X.; Yu, L. A hardware CABAC encoder for HEVC. In Proceedings of the 2013 IEEE International Symposium on Circuits and Systems (ISCAS), Beijing, China, 19–23 May 2013; pp. 1372–1375.
14. Ding, D.; Liu, F.; Qi, H.; Yao, Z. An FPGA-friendly CABAC-encoding architecture with dataflow modeling programming. *Imaging Sci. J.* **2018**, *66*, 346–354. [CrossRef]
15. Wahiba, M.; Abdellah, S.; Aichouche, B.; Azzaz, M. Multiple Bypass Bins FPGA Implementation of H. 256 CABAC Encoder. In Proceedings of the 2018 International Conference on Applied Smart Systems (ICASS), Medea, Algeria, 24–25 November 2018; pp. 1–5.
16. Ramos, F.L.L.; Zatt, B.; Porto, M.S.; Bampi, S. Novel multiple bypass bin scheme and low-power approach for HEVC CABAC binary arithmetic encoder. *J. Integr. Circuits Syst.* **2018**, *13*, 1–11. [CrossRef]
17. Li, W.; Yin, X.; Zeng, X.; Yu, X.; Wang, W.; Fan, Y. A VLSI Implement of CABAC Encoder for H. 265/HEVC. In Proceedings of the 2018 14th IEEE International Conference on Solid-State and Integrated Circuit Technology (ICSICT), Qingdao, China, 31 October–3 November 2018; pp. 1–3.
18. Zhou, D.; Zhou, J.; Fei, W.; Goto, S. Ultra-high-throughput VLSI architecture of H. 265/HEVC CABAC encoder for UHDTV applications. *IEEE Trans. Circuits Syst. Video Technol.* **2014**, *25*, 497–507. [CrossRef]
19. Wahiba, M.; Abdellah, S.; Azzaz, M.S.; Aichouche, B. Design and FPGA Implementation of Residual Data in HEVC CABAC Encoder. In Proceedings of the 2018 International Conference on Signal, Image, Vision and their Applications (SIVA), Guelma, Algeria, 26–27 November 2018; pp. 1–5.
20. Saggiorato, A.V.P.; Ramos, F.L.L.; Zatt, B.; Porto, M.; Bampi, S. HEVC residual syntax elements generation architecture for high-throughput CABAC design. In Proceedings of the 2018 25th IEEE International Conference on Electronics, Circuits and Systems (ICECS), Bordeaux, France, 9–12 December 2018; pp. 193–196.
21. Tran, D.L.; Tran, X.T.; Bui, D.H.; Pham, C.K. An Efficient Hardware Implementation of Residual Data Binarization in HEVC CABAC Encoder. *Electronics* **2020**, *9*, 684. [CrossRef]
22. Nagaraju, M.; Gupta, S.K.; Bhadauria, V. High-throughput, area-efficient hardware architecture of CABAC-Binarization for UHD applications. *Microelectron. J.* **2022**, *123*, 105425. [CrossRef]
23. Vivienne, S.; Madhukar, B.; Gary, J. *High Efficiency Video Coding (HEVC): Algorithms and Architectures*; Springer Publishing Company, Incorporated: Berlin/Heidelberg, Germany, 2014.
24. Fei, W.; Zhou, D.; Goto, S. A 1 gbin/s cabac encoder for H. 264/AVC. In Proceedings of the 2011 19th European Signal Processing Conference, Low-Power HEVC Binarizer Architecture for the CABAC Block Targeting UHD Video Processing, Barcelona, Spain, 29 August–2 September 2011; pp. 1524–12528.
25. Chen, Y.J.; Tsai, C.H.; Chen, L.G. Architecture design of area-efficient SRAM-based multi-symbol arithmetic encoder in H. 264/AVC. In Proceedings of the 2006 IEEE International Symposium on Circuits and Systems (ISCAS), Island of Kos, 21–24 May 2006; pp. 2621–2624.
26. De Matos Alonso, C.; Ramos, F.L.L.; Zatt, B.; Porto, M.; Bampi, S. Low-power HEVC Binarizer architecture for the CABAC block targeting UHD video processing. In Proceedings of the 30th Symposium on Integrated Circuits and Systems Design: Chip on the Sands, Ceara, Brazil, 28 August–1 September 2017; pp. 30–35.
27. Lainema, J.; Ugur, K.; Hallapuro, A. *Single Entropy Coder for HEVC with a High Throughput Binarization Mode*; JCTVC-G569; JCTVC: Geneva, Switzerland, 2011; pp. 1–9.
28. Ramos, F.L.L.; Saggiorato, A.V.P.; Zatt, B.; Porto, M.; Bampi, S. Residual syntax elements analysis and design targeting high-throughput HEVC CABAC. *IEEE Trans. Circuits Syst. Regul. Pap.* **2019**, *67*, 475–488. [CrossRef]

29. Zhang, Y.; Lu, C. Efficient algorithm adaptations and fully parallel hardware architecture of H. 265/HEVC intra encoder. *IEEE Trans. Circuits Syst. Video Technol.* **2018**, *29*, 3415–3429. [CrossRef]
30. Bross, B.; Wang, Y.K.; Ye, Y.; Liu, S.; Chen, J.; Sullivan, G.J.; Ohm, J.R. Overview of the versatile video coding (VVC) standard and its applications. *IEEE Trans. Circuits Syst. Video Technol.* **2021**, *31*, 3736–3764. [CrossRef]

Disclaimer/Publisher's Note: The statements, opinions and data contained in all publications are solely those of the individual author(s) and contributor(s) and not of MDPI and/or the editor(s). MDPI and/or the editor(s) disclaim responsibility for any injury to people or property resulting from any ideas, methods, instructions or products referred to in the content.

Adapting Single-Image Super-Resolution Models to Video Super-Resolution: A Plug-and-Play Approach

Wenhao Wang, Zhenbing Liu *, Haoxiang Lu, Rushi Lan and Yingxin Huang

School of Computer Science and Information Security, Guilin University of Electronic Technology, Guilin 541004, China
* Correspondence: zbliu@guet.edu.cn

Abstract: The quality of videos varies due to the different capabilities of sensors. Video super-resolution (VSR) is a technology that improves the quality of captured video. However, the development of a VSR model is very costly. In this paper, we present a novel approach for adapting single-image super-resolution (SISR) models to the VSR task. To achieve this, we first summarize a common architecture of SISR models and perform a formal analysis of adaptation. Then, we propose an adaptation method that incorporates a plug-and-play temporal feature extraction module into existing SISR models. The proposed temporal feature extraction module consists of three submodules: offset estimation, spatial aggregation, and temporal aggregation. In the spatial aggregation submodule, the features obtained from the SISR model are aligned to the center frame based on the offset estimation results. The aligned features are fused in the temporal aggregation submodule. Finally, the fused temporal feature is fed to the SISR model for reconstruction. To evaluate the effectiveness of our method, we adapt five representative SISR models and evaluate these models on two popular benchmarks. The experiment results show the proposed method is effective on different SISR models. In particular, on the Vid4 benchmark, the VSR-adapted models achieve at least 1.26 dB and 0.067 improvement over the original SISR models in terms of PSNR and SSIM metrics, respectively. Additionally, these VSR-adapted models achieve better performance than the state-of-the-art VSR models.

Keywords: video super-resolution; single-image super-resolution; plug-and-play; deformable convolution

1. Introduction

Numerous videos are captured every day; however, due to the different capabilities of sensors, the quality of captured videos can vary greatly, which affects the subsequent analysis and applications [1–4]. Recently, computer technologies have been applied to many fields [5–8]. In particular, video super-resolution (VSR) is a technology for improving the quality of captured video. It produces high-resolution (HR) video frames from their low-resolution (LR) counterparts. The VSR problem is challenging due to its ill-posed nature, but its applications include video display, video surveillance, video conferencing, and entertainment [9].

VSR models take consecutive frames as input. Single-image super-resolution (SISR) methods process only one image at a time. So, VSR models take both spatial information and temporal information into account, while SISR models only exploit spatial information for super-resolution (SR) reconstruction. Thus, many VSR methods adapt SISR models for spatial information extraction. For example, Haris et al. [10] introduced RBPN, which employs blocks from DBPN [11] in a recurrent encoder–decoder module to utilize spatial and temporal information. Tian et al. [12] adapted EDSR [13] as the main design for the SR reconstruction network in TDAN. Liang et al. [14] utilized residual Swin Transformer blocks from SwinIR [15] in their proposed RVRT. Although these works have adapted SISR models, each method utilizes only one SISR model. Applying SISR techniques to the

VSR models would require considerable effort and they may not perform as effectively as specialized VSR models.

Meanwhile, several VSR methods do not rely on SISR models. For instance, Xue et al. [16] proposed TOF, which estimates task-oriented flow to recover details in SR frames. Wang et al. [17] proposed SOF-VSR, which estimates HR optical flow from LR frames. SWRN [18] can be utilized in real time on a mobile device. However, the development of a VSR model without adapting SISR methods is very costly, as the model needs to capture both temporal and spatial information. Moreover, compared with SISR methods, they may be less effective in utilizing spatial information.

To alleviate the above issues, we propose a plug-and-play approach for adapting existing SISR models to the VSR task. Firstly, we summarize a common architecture of SISR models and provide a formal analysis of adaptation to achieve better effectiveness of different SISR models. Then, we present an adaptation method, which inserts a plug-and-play temporal feature extraction module into SISR models. Specifically, the temporal feature extraction module consists of three submodules. The spatial aggregation submodule aligns features extracted by the original SISR model. The alignment is performed based on the result of the offset estimation submodule. Then, the temporal aggregation submodule is applied to aggregate information extracted from all neighboring frames.

To evaluate the effectiveness of the proposed method, we adapt five representative SISR models, i.e., SRResNet [19], EDSR [13], RCAN [20], RDN [21], and SwinIR [15], and the evaluations are conducted on two popular benchmarks, i.e., Vid4 and SPMC-11. On the Vid4 benchmark, the VSR-adapted models achieve at least 1.26 dB and 0.067 improvements over original SISR models in terms of peak signal-to-noise ratio (PSNR) [22] and structural similarity index (SSIM) [23], respectively. On the SPMC benchmark, the VSR-adapted models achieve at least 1.16 dB and 0.036 gain over original SISR models in terms of PSNR and SSIM, respectively. Moreover, the VSR-adapted models surpassed the performance of state-of-the-art VSR models.

For this paper, the main contributions are as follows: (1) We propose a plug-and-play approach for adapting SISR models to the VSR task. Instead of adapting one SISR model, the proposed method is based on a common architecture of SISR models. (2) A plug-and-play temporal feature extraction module is introduced. Thus, the adapted model gains the capability to exploit temporal information. (3) Extensive experiments are conducted to evaluate its effectiveness.

2. Related Work

2.1. Single-Image Super-Resolution

The SISR problem is an ill-posed problem, and learning-based methods have significantly improved the performance in terms of accuracy [13,15,19–21,24,25] and speed [26–29]. In 2014, Dong et al. [30] introduced a learning-based model, namely SRCNN, into the SISR field. Inspired by ResNet [31], Ledig et al. [19] proposed SRResNet in 2017. SRResNet [19] accepts LR images directly and achieves high performance and increased efficiency. Kim et al. [13] improved the SRResNet by removing unnecessary batch normalization in residual blocks and expanding the number of parameters. In 2018, Zhang et al. [21] employed a densely connected architecture. All extracted features are fused to utilize hierarchical information. Subsequently, Zhang et al. [20] introduced the channel attention mechanism that adaptively weights features channel-wisely. In 2021, Liang et al. [15] proposed SwinIR by making use of the Transformer [32]. Additionally, SwinIR uses the Swin Transformer [33] variation, which is more appropriate for computer vision tasks. By appropriately employing convolution layers and Swin Transformer modules, SwinIR can capture local and global dependencies at the same time, resulting in SOTA performance.

2.2. Video Super-Resolution

In recent years, deep-learning-based models have been used to solve the VSR problem, and have become increasingly popular [9]. We roughly divide VSR models into two categories:

(1) Models adapting SISR models: Sajjadi et al. [34] proposed FRVSR, which takes EnhanceNet [35] as the subnetwork for SR reconstruction. Haris et al. [10] applied the iterative up- and downsampling technique [11] in RBPN. The representative deep learning SISR model, EDSR [13], is utilized by many VSR models. Tian et al. [12] applied a shallow version of EDSR [13] in TDAN. EDVR [36] and WAEN [37] both employed the residual block and upsampling module from EDSR [13] in the reconstruction module. Inspired by [12], Xu et al. [38] adapted EDSR as the reconstruction module. EGVSR [39] applied ESPCN [26] as the backbone for the SR net. The recently proposed RVRT [14] utilized the residual Swin Transformer block, which is proposed in SwinIR [15].

(2) Models without adapting SISR models: DUF [40] reconstructs SR frames by estimating upsampling filters and a residual image for high-frequency details. Kim et al. [41] employed 3D convolution to capture spatial–temporal nonlinear characteristics between LR and HR frames. Xue et al. [16] proposed a method, namely TOF. It learns a task-specific representation of motion. Wang et al. [17] proposed SOF-VSR, which estimates HR optical flow from LR frames. To better leverage the temporal information, TGA [42] introduced a hierarchical architecture. Recently, Chan et al. [43] proposed BasicVSR by investigating the essential components of VSR models. Liu et al. [44] applied spatial convolution packing to jointly exploit spatial–temporal features. For better fusing information from neighboring frames, Lee et al. [45] utilized both attention-based alignment and dilation-based alignment. Lian et al. [18] proposed SWRN to achieve real-time inference while producing superior performance.

Because VSR models have to capture both temporal and spatial information, proposing a VSR method requires more effort. Thus, many researchers turn to adapting SISR models. Based on SISR models, proposing a VSR method can focus on capturing temporal information. However, these models either utilize a SISR model as a subnet or adapt modules from a SISR model to extract features. Additionally, they may be less effective than those methods that do not adapt SISR methods. Our work proposed a plug-and-play approach to adapt SISR models to the VSR task. The proposed method works on different SISR models as it follows the common architecture of SISR models we have summarized. The spatial information and temporal information are both extracted in the proposed method.

3. Methodology

In this section, we first summarize the common architecture of SISR models. Then, we provide a formal analysis of adaptation. Following that, a general VSR adaptation method is proposed. Finally, we present a plug-and-play temporal feature extraction module.

3.1. Revisit of Single-Image Super-Resolution Models

For the effectiveness on different SISR models [13,15,19–21,46], we first summarize a common architecture, as shown in Figure 1. For simplicity, some operations such as element-wise addition and concatenation are omitted. As shown in Figure 1a, the common architecture of SISR models can be divided into three modules: shallow feature extraction (FE) module, deep FE module, and reconstruction module. Figure 1b–e illustrate the details of four SISR models. As one can see, the shallow FE module takes one LR image as input and extracts features by a few convolution layers. The deep FE module consists of several submodules or blocks, where advanced techniques, such as dense connection [21], channel attention [20], and self-attention [15], are applied. Thus, the deep FE module is where the key novelty of SISR models lies. Finally, the features from the deep FE module are fed to the reconstruction module to produce the SR image.

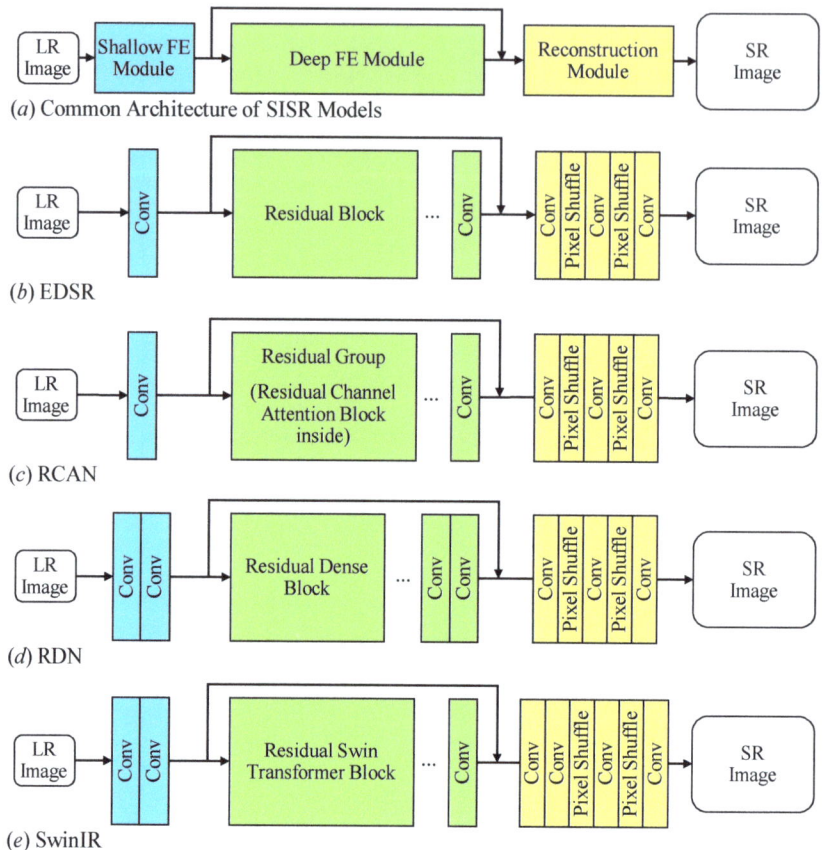

Figure 1. The architectures of typical SISR models.

Thus, given an LR image $\mathbf{y} \in \mathbb{R}^{H \times W \times 3}$, these SISR models can be generalized using the following representation:

$$\mathbf{x} = Method_{SISR}(\mathbf{y}), \tag{1}$$

where $Method_{SISR}(\cdot)$ is the SISR model. $\mathbf{x} \in \mathbb{R}^{sH \times sW \times 3}$ represents the SR result with upscale factor s. H and W denote the height and width of LR image, respectively. According to the common architecture of SISR models, Equation (1) can be expanded as

$$\mathbf{x} = Recons(FE_{deep}(FE_{shallow}(\mathbf{y})) + FE_{shallow}(\mathbf{y})), \tag{2}$$

where the shallow and deep FE modules are noted as $FE_{shallow}(\cdot)$ and $FE_{deep}(\cdot)$, respectively. The reconstruction module is denoted as $Recons(\cdot)$.

Different from the SISR problem, the VSR methods have to exploit both spatial and temporal information. Thus, we make use of sliding window framework [12] to capture temporal dependency. Given consecutive $2n + 1$ LR frames $\mathbf{Y} = \{\mathbf{y}_{t-n}, \cdots \mathbf{y}_{t-1}, \mathbf{y}_t, \mathbf{y}_{t+1}, \cdots \mathbf{y}_{t+n}\}$, the representation of VSR models is formulated as

$$\mathbf{x}_t = Method_{VSR}(\mathbf{Y}), \tag{3}$$

where the VSR method is $Method_{VSR}(\cdot)$. $\mathbf{x}_t$ represents the reconstructed SR frame, the frame index of which is t.

Note that the main difference between Equations (1) and (3) is the input, and Equation (2) is an expanded representation of Equation (1). In order to adapt existing SISR models to the VSR task, a straightforward method is to modify the shallow FE module. Then, the adapted model can be represented as

$$x_t = Recons(FE_{deep}(FE'_{shallow}(\mathbf{Y})) + FE'_{shallow}(\mathbf{Y})), \quad (4)$$

where $FE'_{shallow}(\cdot)$ is the modified shallow FE module.

3.2. Proposed Video Super-Resolution Adaptation Method

According to the analysis in Section 3.1, we propose a general method to easily adapt SISR models to the VSR task. As shown in Figure 2, the architecture of the proposed VSR-adapted models consists of 4 modules. Firstly, the VSR-adapted model applies the shallow FE module $FE_{shallow}(\cdot)$ to obtain low-level features $\mathbf{F}_{s,i} \in \mathbb{R}^{H \times W \times C}$ for each LR frame $\mathbf{y}_i$. The subscript i represents the relative index of the center frame. The center frame is denoted as $_0$, and C stands for the number of channels in a feature. The shallow feature of center frame $\mathbf{F}_{s,0}$ is skip-connected to the output of the deep FE module with element-wise addition for global residual leaning. Secondly, the temporal FE module $FE_{temporal}(\cdot)$ is employed to exploit spatial–temporal information. It takes LR frames to estimate the offsets of pixels. It also takes shallow features which will be spatially aggregated based on the offsets. In order to enable the deep FE module to leverage information from all LR frames, spatial-aggregated features are temporally aggregated in the temporal FE module. Thirdly, the deep FE module $FE_{deep}(\cdot)$ is responsible for estimating accurate residual features with advanced techniques. Finally, the reconstruction module $Recons(\cdot)$ upsamples features with specific scale factors and produces SR frames. The architecture can be represented as

$$\mathbf{F}_{s,i} = FE_{shallow}(\mathbf{y}_i), \quad (5)$$

$$\mathbf{F}_T = FE_{temporal}(\mathbf{F}_{s,-n}, \cdots, \mathbf{F}_{s,0}, \cdots, \mathbf{F}_{s,n}, \mathbf{y}_{-n}, \cdots, \mathbf{y}_0, \cdots, \mathbf{y}_n), \quad (6)$$

$$x_0 = Recons(FE_{deep}(\mathbf{F}_T) + \mathbf{F}_{s,0}), \quad (7)$$

where i denotes the relative index of the target frame, ranging from $-n$ to n. The temporal feature $\mathbf{F}_T \in \mathbb{R}^{H \times W \times C}$ is the output of temporal FE module.

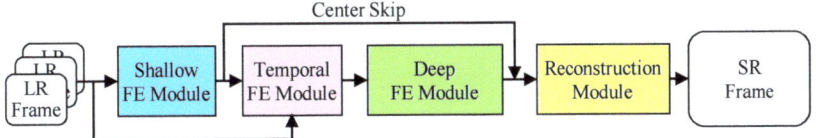

Figure 2. The Architecture of Proposed General VSR-Adapted Models.

For adapting different SISR models, the proposed method maintains the shallow FE module, deep FE module, and reconstruction module unmodified. Furthermore, we employ the temporal feature extraction module between the shallow FE module and the deep FE module in accordance with accuracy and latency concerns.

From an accuracy perspective, the main difference between an input LR frame and its ground truth HR frame is the high-frequency content. Thus, the better the residual feature that is extracted, the better the achieved performance. The proposed architecture takes advantage of the deep FE module, where the key novelties of SISR models lie [46]. Further, with the information from neighboring frames, the deep FE module is able to extract more accurate features for reconstruction. Thus, the temporal FE module is employed before deep FE module.

From a latency perspective, the temporal FE module aggregates the features extracted from all input frames. It requires previous modules to complete their processing for each frame. To minimize the overall computation time, the proposed temporal FE module is employed after shallow FE module because its relatively small number of layers has a negligible impact on inference latency.

3.3. Plug-and-Play Temporal Feature Extraction Module

In order to exploit spatial–temporal information, the temporal FE module is proposed. The detailed architecture is illustrated in Figure 3, which consists of three submodules, i.e., offset estimation, spatial aggregation, and temporal aggregation.

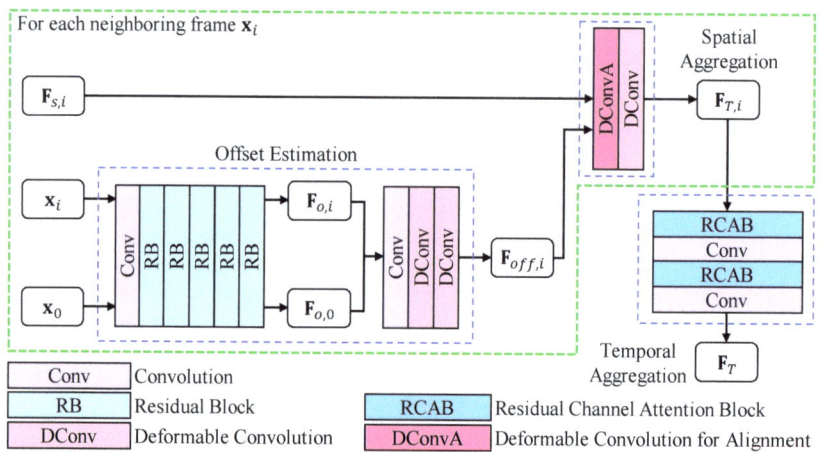

Figure 3. The Temporal Feature Extraction Module.

The offset estimation submodule takes the center LR frame y_0 and each neighboring frame y_i as inputs. The intermediate feature extraction is performed by a convolution layer and five residual blocks, and the parameters are shared across all input LR frames. The intermediate features are noted as $\mathbf{F}_{o,i} \in \mathbb{R}^{H \times W \times C}$. The offset feature $\mathbf{F}_{off,i} \in \mathbb{R}^{H \times W \times C}$ is estimated from the intermediate feature $\mathbf{F}_{o,0}$ and $\mathbf{F}_{o,i}$ using a convolution layer and two deformable convolution layers. The offset estimation submodule can be formulated as

$$\mathbf{F}_{o,i} = RB_5(\cdots RB_1(Conv_1(\mathbf{y}_i))\cdots), \tag{8}$$

$$\mathbf{F}_{off,i} = DConv_2(DConv_1(Conv_2(CAT(\mathbf{F}_{o,i}, \mathbf{F}_{o,0})))), \tag{9}$$

where $RB(\cdot)$ is residual block. $Conv(\cdot)$ and $DConv(\cdot)$ are convolution and deformable convolution, respectively. The concatenation is denoted as $CAT(\cdot)$.

The shallow feature $\mathbf{F}_{s,i}$ and the estimated offset $\mathbf{F}_{off,i}$ are then fed into the spatial aggregation submodule. Here, a variation of deformable convolution is used to extract features $\mathbf{F}_{s,i}$, which takes $\mathbf{F}_{off,i}$ for offset. This allows the offset feature $\mathbf{F}_{off,i}$ to guide the alignment in the spatial aggregation submodule. Another deformable convolution is applied for refinement, resulting in output feature $\mathbf{F}_{T,i} \in \mathbb{R}^{H \times W \times C}$. The spatial aggregation submodule can be given by

$$\mathbf{F}_{T,i} = DConv_3(DConvA(\mathbf{F}_{s,i}, \mathbf{F}_{off,i})), \tag{10}$$

where $DConvA(\cdot, \cdot)$ is the variation of deformable convolution. The variation of deformable convolution $DConvA(\cdot, \cdot)$ takes the first input for feature extraction and the second input for offset.

After spatial aggregation, the temporal aggregation submodule fuses these spatial-aggregated features $\mathbf{F}_{T,-n} \cdots \mathbf{F}_{T,n}$. For fusing a feature with $(2n+1) \times C$ channels, a simple convolution layer is not sufficient. Therefore, a residual channel attention block [20] is employed to adaptively weight these features channel-wise. A convolution layer for channel reduction is then applied. The channel shrinkage is performed in two steps to minimize information loss: first reducing to twice the SISR features' channels and then reducing to once. The temporal aggregation submodule can be represented as

$$\mathbf{F}_T = Conv_4(RCAB_2(Conv_3(RCAB_1(CAT(\mathbf{F}_{T,-n}, \cdots, \mathbf{F}_{T,n}))))), \quad (11)$$

where $RCAB_1(\cdot)$ and $RCAB_2(\cdot)$ are residual channel attention blocks. The number of channels of the features output by $Conv_3(\cdot)$ and $Conv_4(\cdot)$ is $2 \times C$ and C, respectively. The temporal-aggregated feature is $\mathbf{F}_T \in \mathbb{R}^{H \times W \times C}$.

Overall, the spatial aggregation aligns neighboring features based on the result of the offset estimation submodule. Then, the temporal aggregation submodule fuses the spatial-aggregated features, resulting in an output containing information from all input LR frames. Finally, the plug-and-play module extracts feature $\mathbf{F}_T$, which contains spatial–temporal information from all input frames. Further, we summarize the detailed algorithm of the VSR-adapted method with plug-and-play temporal feature extraction module in Algorithm 1. For easy understanding, we divided the loop into multiple ones.

Algorithm 1: Video Super-Resolution with SISR Model and Plug-and-Play Temporal Feature Extraction Module.

Input: Consecutive low-resolution frames $\mathbf{y}_i$. i is relative index to the center frame ranging from $-n$ to n.
Output: Super-resolution center frame $\mathbf{x}_0$.
// Shallow FE module from SISR model
1 **for** $i = -n, -n+1, \cdots, n$ **do**
2 $\quad$ $\mathbf{F}_{s,i} = FE_{shallow}(\mathbf{y}_i)$;
3 **end**
// Offset estimation submodule of temporal FE module
4 **for** $i = -n, -n+1, \cdots, n$ **do**
5 $\quad$ $\mathbf{F}_{o,i} = RB_5(\cdots RB_1(Conv_1(\mathbf{y}_i))\cdots)$;
6 $\quad$ $\mathbf{F}_{off,i} = DConv_2(DConv_1(Conv_2(CAT(\mathbf{F}_{o,i}, \mathbf{F}_{o,0}))))$;
7 **end**
// Spatial aggregation submodule of temporal FE module
8 **for** $i = -n, -n+1, \cdots, n$ **do**
9 $\quad$ $\mathbf{F}_{T,i} = DConv_3(DConvA(\mathbf{F}_{s,i}, \mathbf{F}_{off,i}))$;
10 **end**
// Temporal aggregation submodule of temporal FE module
11 $\mathbf{F}_T = Conv_4(RCAB_2(Conv_3(RCAB_1(CAT(\mathbf{F}_{T,-n}, \mathbf{F}_{T,-n+1}, \cdots, \mathbf{F}_{T,n})))))$;
// Deep FE module and reconstruction module from SISR model
12 $\mathbf{x}_0 = Recons(FE_{deep}(\mathbf{F}_T) + \mathbf{F}_{s,0})$;

---

4. Experiment

4.1. Datasets

Following previous studies [12,16,47], we utilized the widely used Vimeo90K dataset for training. This dataset includes videos with different scenarios, such as moving objects, camera motion, and complex scene structures. It consists of 90,000 video clips with a resolution of 448×256. As per the official split, we use 64,612 video clips for training. The HR frames of these videos were used as the ground truth. For training, we randomly cropped these HR frames to patches with the size of 256×256, and these patches were bicubically downsampled to the size of 64×64 using the Matlab function *imresize*. We randomly flipped and rotated the data during training.

For testing, we evaluated the effectiveness of our proposed model on two public benchmarks, i.e., the Vid4 [48] and SPMC-11 [47]. The quantitative metrics were PSNR [22] and SSIM [23], computed in the luminance (Y) channel. We also cropped 8 pixels near the image boundary, similar to the previous approach [12].

4.2. Implementation Details

To evaluate the proposed method, we employed it on five representative SISR models: (1) SRResNet [19] is the generator model in SRGAN. (2) EDSR [13] is a representative SISR model. (3) RCAN [20] makes use of channel attention. (4) RDN [21] has the advantage of a dense connection. (5) SwinIR [15] introduces Swin Transformer [33]. For SISR models, we generated SR videos frame by frame.

In our implementation of SRResNet [19], we removed all batch norm layers. We used the EDSR baseline [13] with a feature channel count and block count of 64 and 16, respectively. For SwnIR [15], the LR patch size was 48×48, and the GT patch size was 192×192. We used a smaller patch size for SwinIR for lower memory consumption. The batch size for training all models was 16. We empirically set $n = 2$, indicating that a VSR-adapted model takes five frames as input. For SISR models, the number of input frames was one. Each SISR model and its VSR-adapted model were trained from scratch using the same setting except for the number of input frames.

We used the mean square error (MSE) as the loss function, defined as $Loss = \|HR - SR\|^2$. The parameters were updated using the Adam optimizer [49] with $\beta 1 = 0.9$ and $\beta 2 = 0.99$. The learning rate was initialized as 1×10^{-4} and halved for every 1×10^5 iterations. We trained the models for 3×10^5 iterations. All experiments were implemented in Pytorch and ran on a server with NVIDIA GPUs.

4.3. Effectiveness on Different Single-Image Super-Resolution Models

To evaluate the effectiveness of the proposed method, we conducted experiments on five representative SISR models. Table 1 displays the quantitative results on two popular benchmarks. The PSNR and SSIM metrics of VSR-adapted models improved by at least 1.16 dB and 0.036, respectively. It demonstrates that the proposed method works effectively on various SISR models. Moreover, the performance of the VSR-adapted models is positively correlated with the capacity of the original models. In the SISR task, EDSR [13] is better than SRResNet [19] but underperforms RCAN [20] and RDN [21]. The performance of RCAN and RDN is on par, and SwinIR [15] has the best performance. As shown in Table 1, the VSR-adapted models exhibit similar trends. We use the suffix "-VSR" to represent the VSR-adapted models. The performances of SRResNet-VSR and EDSR-VSR are weaker than those of RCAN-VSR and RDN-VSR, and SwinIR-VSR achieves the best results on both benchmarks. Moreover, we computed the PSNR metric on the Vid4 benchmark during training. As illustrated in Figure 4, the VSR-adapted models benefit from the information aggregated from neighboring frames, and they performed better in the early iterations during training. Thus, the proposed method is effective on different SISR models, and the plug-and-play temporal feature extraction module enables the VSR-adapted models to exploit spatial and temporal information.

Further, we visualized the results of the Vid4 and SPMC-11 benchmarks for qualitative comparison. Several processed frames are shown in Figures 5 and 6. We can observe that the VSR-adapted models provide visually appealing results. By contrast, the original SISR models produce blurry SR frames and incorrect textures. Overall, the VSR-adapted models reconstruct results with clearer text, richer textures, and fewer artifacts. Among the results of the VSR-adapted models, SRResNet-VSR and EDSR-VSR produce more artifacts than other VSR-adapted models. This is consistent with the capabilities of original SISR models.

Table 1. Quantitative Comparison of SISR Models and VSR-Adapted Models on Vid4 and SPMC-11. The best results are in bold.

Benchmark	Method	Original		VSR Adapted	
		PSNR	SSIM	PSNR	SSIM
Vid4	SRResNet [19]	25.30	0.728	26.56	0.797
	EDSR [13]	25.27	0.726	26.58	0.798
	RCAN [20]	**25.45**	0.737	26.74	0.804
	RDN [21]	25.40	0.734	26.75	0.806
	SwinIR [15]	25.41	**0.738**	**26.84**	**0.811**
SPMC-11	SRResNet [19]	27.92	0.815	29.16	0.853
	EDSR [13]	27.85	0.813	29.14	0.853
	RCAN [20]	28.32	0.823	29.48	0.859
	RDN [21]	28.24	0.821	29.55	0.862
	SwinIR [15]	**28.46**	**0.826**	**29.74**	**0.866**

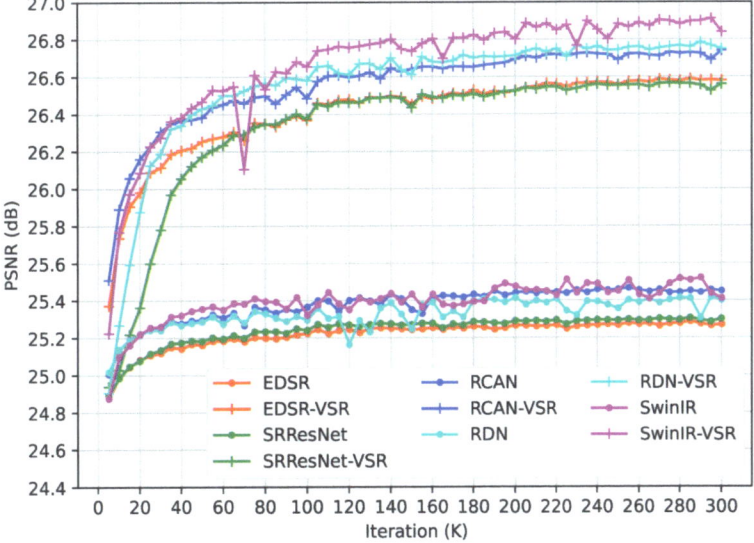

Figure 4. The PSNR Curve on Vid4 Benchmark During Training.

4.4. Comparisons with State-of-the-Art Methods

We compared these VSR-adapted models with 10 state-of-the-art VSR algorithms, i.e., STAN [50], EGVSR [39], TOFlow [16], STMN [51], SOF-VSR [17], ST-CNN [44], TDAN [12], D3Dnet [47], FRVSR [34], and WAEN [37]. Table 2 shows the quantitative metrics on the Vid4 and SPMC-11 benchmarks. The values with † are reported in [47]. As shown in Table 2, the VSR-adapted models achieve competitive performance on both Vid4 and SPMC-11 benchmarks. All VSR-adapted models perform better than D3Dnet. Compared with D3Dnet, the SRResNet-VSR and EDSR-VSR achieve comparative performance. The performances achieved by RCAN-VSR and RDN-VSR are between FRVSR and WAEN. Among them, the SwinIR-VSR outperforms all models in terms of PSNR metrics.

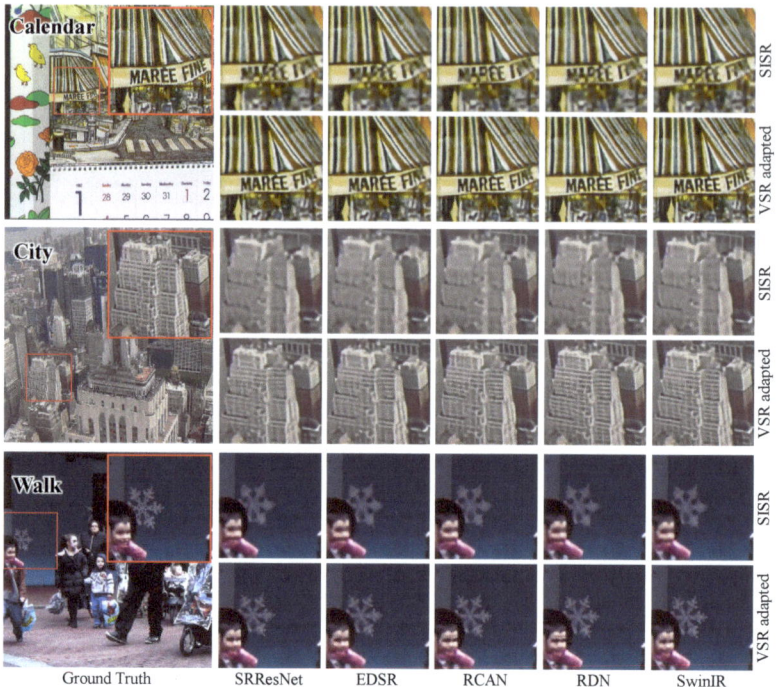

Figure 5. The Qualitative Comparison of SISR Models and Corresponding VSR Adaptations on Vid4 Benchmark.

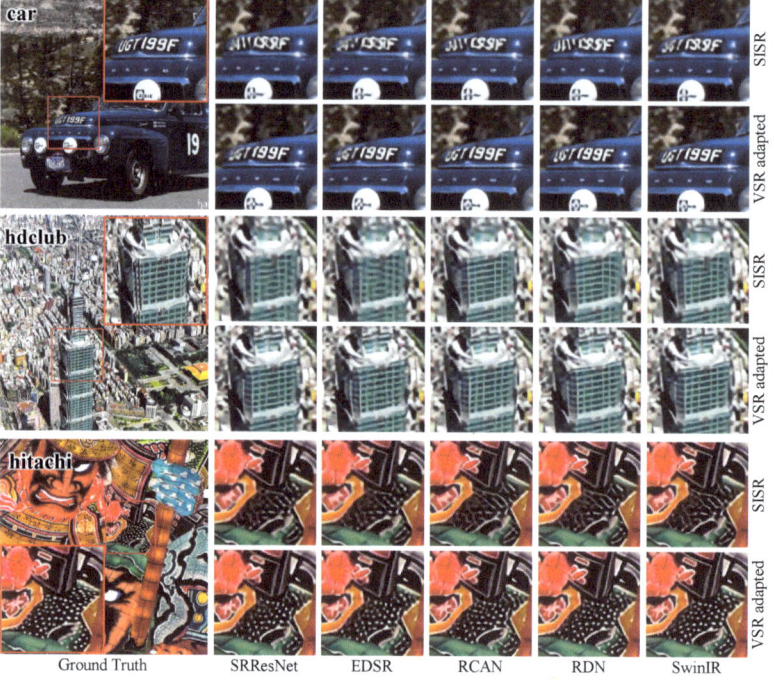

Figure 6. The Qualitative Comparison of SISR Models and Corresponding VSR Adaptations on SPMC-11 Benchmark.

Table 2. Quantitative comparison of Vid4 and SPMC-11. The best results are in bold. The values with † are reported in [47].

Method	Vid4 PSNR (dB)	SSIM	SPMC-11 PSNR (dB)	SSIM
STAN [50]	25.58	0.743	—	—
EGVSR [39]	25.88	0.800	—	—
TOFlow [16]	25.90	0.765	—	—
STMN [51]	25.90	0.788	—	—
SOF-VSR [17]	26.02	0.772	28.21 †	0.832 †
ST-CNN [44]	26.12	**0.823**	—	—
TDAN [12]	26.42	0.789	28.51 †	0.841 †
D3Dnet [47]	26.52	0.799	28.78	0.851
FRVSR [34]	26.69	0.822	—	—
WAEN [37]	26.79	—	—	—
SRResNet-VSR	26.56	0.797	29.16	0.853
EDSR-VSR	26.58	0.798	29.14	0.853
RCAN-VSR	26.74	0.804	29.48	0.859
RDN-VSR	26.75	0.806	29.55	0.862
SwinIR-VSR	**26.84**	0.811	**29.74**	**0.866**

For a finer quantitative comparison on the Vid4 benchmark, we illustrate the PSNR metric of each frame in Figure 7. For simplicity, we select four models, i.e., TDAN [12], FRVSR [34], EDSR-VSR, and SwinIR-VSR. Compared with TDAN, the EDSR-VSR achieves similar performance. Note that the first two and last two frames show a greater difference between TDAN and EDSR-VSR. Because there is less neighboring information for VSR models to exploit, the VSR models exhibit poor performance at the beginning and end of a video. Compared with FRVSR, the SwinIR-VSR achieved better performance on the *Calendar* and *Walk*. As the frame index increases on the *Calendar*, the gap between SwinIR-VSR and FRVSR becomes smaller. Additionally, the performance of SwinIR-VSR is lower than that of FRVSR after the first five frames on the *City*. This is because the SwinIR-VSR makes use of neighboring frames in a sliding window scheme while the FRVSR utilizes them in a recurrent scheme.

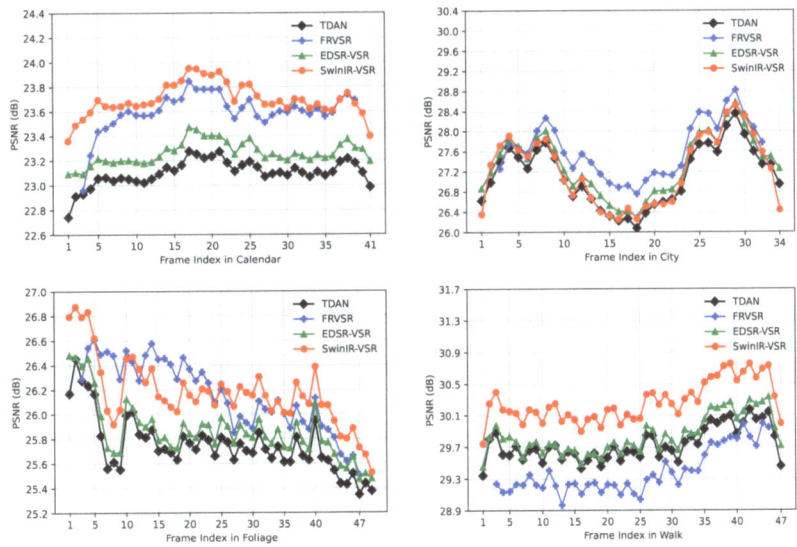

Figure 7. The PSNR curve of VSR models on Vid4 benchmark.

For a qualitative comparison, we compared the VSR-adapted models to SOF-VSR [17], TOF [16], TDAN [12], D3Dnet [47], and FRVSR [34]. As shown in Figure 8, the VSR-adapted models reconstruct visually attractive results. The text on the *Calendar* is now easier to read and the details of the *City* are clearer. Additionally, the clothes in the *Walk* image are more recognizable. Moreover, we observed similar trends in the SPMC-11 benchmark, as illustrated in Figure 9. The quality of the reconstructed results of EDSR-VSR is equivalent to that of the compared methods. The RDN-VSR and RCAN-VSR provide results with better quality. The result of SwinIR-VSR has the least artifacts.

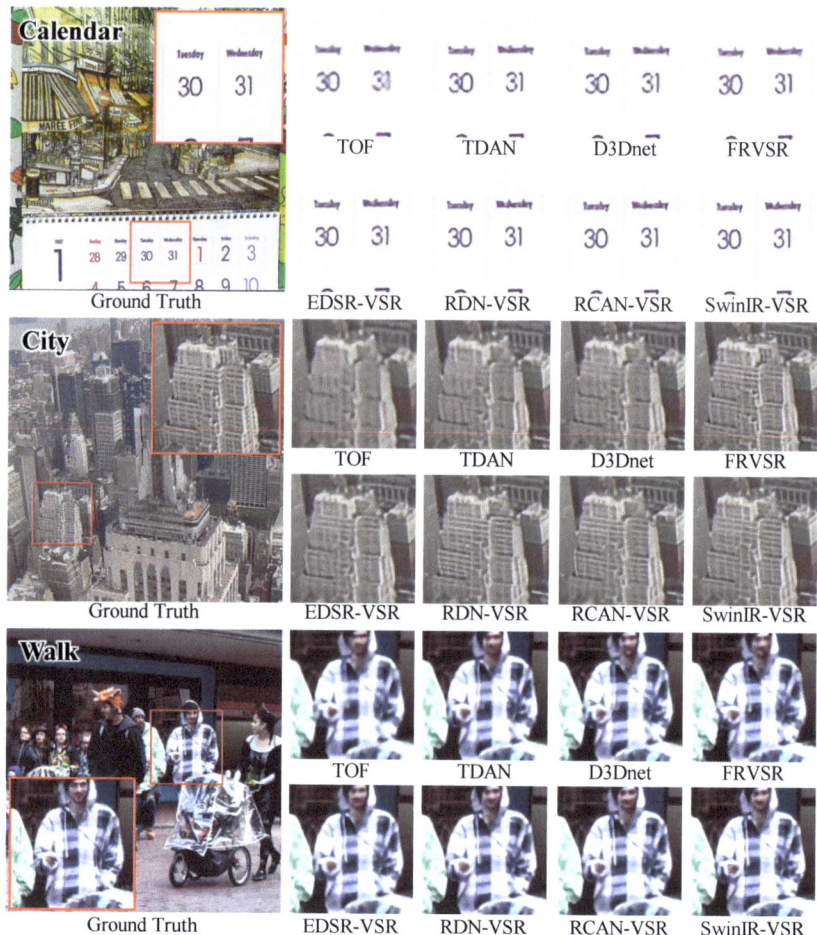

Figure 8. Qualitative Comparison of VSR Models on Vid4 Benchmark.

4.5. Comparisons of Temporal Consistency

To evaluate the temporal consistency of the proposed method, we generated temporal profiles according to [34] for visualization. As shown in Figure 10, the positions of temporal profiles are highlighted with red lines. The heights of temporal profiles vary due to the video length. As shown in the *Calendar*, the temporal profiles demonstrate that the original SISR models perform poorly because they are unable to capture temporal information. By contrast, the VSR methods and VSR-adapted models produce results with fewer artifacts. However, inappropriate aggregation of temporal information can lead to degraded results. As illustrated in the *City*, the original SISR models and our VSR-adapted models exhibit better temporal consistency than VSR models.

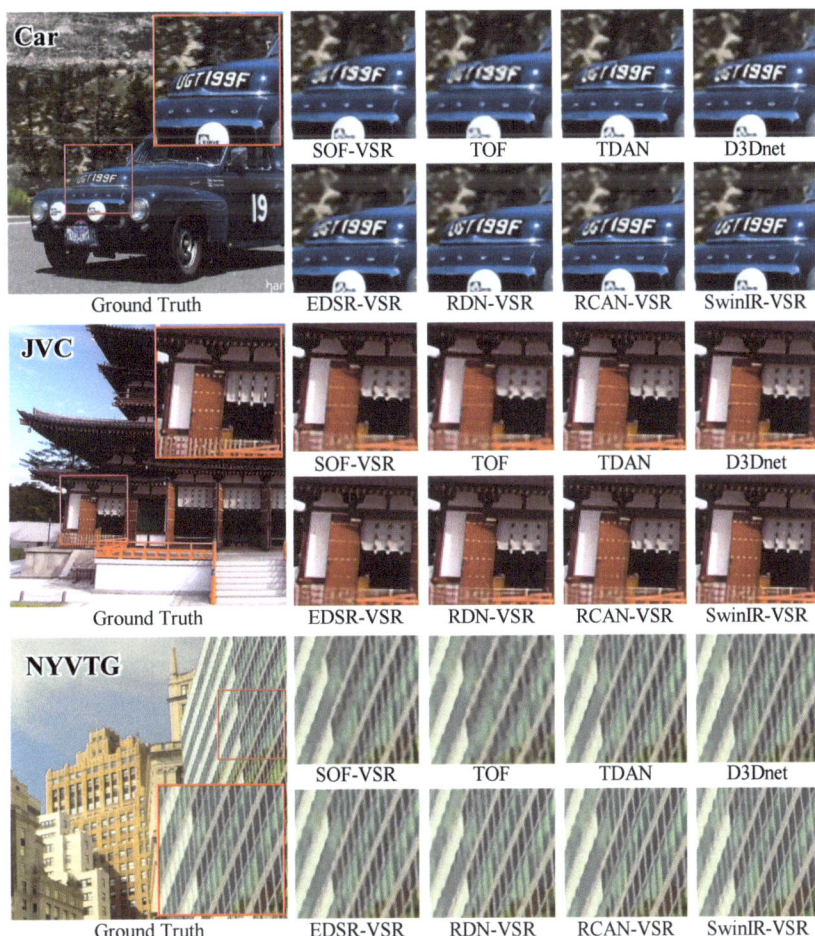

Figure 9. Qualitative Comparison of VSR Models on SPMC-11 Benchmark.

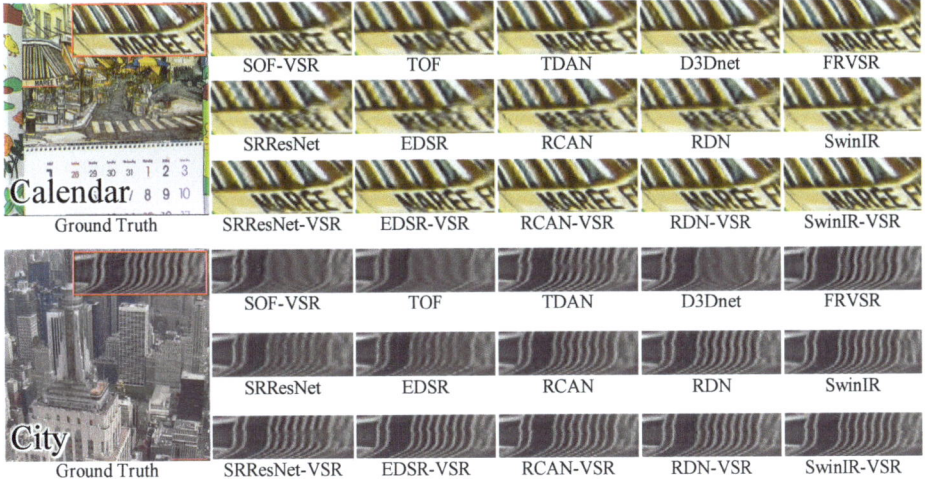

Figure 10. Qualitative Comparison of Temporal Profile on Vid4 Benchmark.

4.6. Ablation Study

We used EDSR [13] as the baseline in the ablation study to evaluate the effectiveness of the proposed temporal feature extraction module, which consists of offset estimation, spatial aggregation, and temporal aggregation submodules. We evaluated three models to determine the effectiveness of each submodule. The first variation is denoted as Model 1. We fed shallow features from neighboring frames to the spatial aggregation submodule without the support of the offset estimation submodule. The neighboring features were then fused with a convolution using a 1×1 kernel. Model 2 is referred to as the second variation. We introduced the offset estimation submodule, which makes use of the center frame and neighboring frames to guide the spatial aggregation. The third variation, denoted as EDSR-VSR, combines all the components, including channel attention and progressive channel shrinking.

Table 3 indicates that relying solely on the spatial aggregation submodule does not lead to performance improvement. However, with the support of the offset estimation submodule, there is a significant performance improvement. Furthermore, the temporal aggregation submodule further improved the performance. Three submodules play an irreplaceable role in our presented temporal feature extraction module.

Table 3. The Effectiveness of Each Component in Temporal Feature Extraction Module.

Dataset	Model	Spatial Aggregation	Offset Estimation	Temporal Aggregation	PSNR (dB)	SSIM
Vid4	EDSR [13]	✗	✗	✗	25.27	0.726
	Model 1	✓	✗	✗	25.31	0.725
	Model 2	✓	✓	✗	26.49	0.793
	EDSR-VSR	✓	✓	✓	26.58	0.798
SPMC-11	EDSR [13]	✗	✗	✗	27.85	0.813
	Model 1	✓	✗	✗	27.88	0.813
	Model 2	✓	✓	✗	28.97	0.849
	EDSR-VSR	✓	✓	✓	29.14	0.853

To evaluate the efficiency of the proposed method, we conducted a comparison on the Vid4 benchmark. We evaluated three models, i.e., EDSR [13], EDSR-VSR, and EDSR-VSR 2. The EDSR-VSR 2 employs the temporal feature extraction module after the deep feature extraction module. Table 4 shows the performance and average latency of inference. As we can see, the EDSR-VSR is about 1.6× faster than the EDSR-VSR 2. Although the EDSR-VSR is slower than EDSR [13], it reaches 24 frames per second. Specifically, we analyzed the latency of each part of EDSR-VSR. Overall, 0.89% of the latency is consumed by the shallow feature extraction module from the SISR model. The subsequent offset estimation submodule, spatial aggregation submodule, and temporal aggregation submodule occupied 21.25%, 39.99%, and 15.21% of the latency, respectively. Additionally, 22.66% of the time is spent on the deep feature extraction and reconstruction module from the SISR model. Note that the temporal feature extraction module has to process all input frames, so each submodule takes a longer time to complete the computation. Thus, the proposed method balances the accuracy and latency.

Table 4. The Efficiency of Proposed Method on Vid4 Benchmark.

	EDSR [13]	EDSR-VSR	EDSR-VSR 2
PSNR (dB)	25.27	26.58	26.61
SSIM	0.726	0.798	0.798
Latency (ms)	9.872	41.543	65.003

5. Discussion and Limitation

The proposed method builds a bridge between the SISR model and the VSR model. We revisited many SISR models and summarized a common architecture of SISR models. The proposed method leverages the inherent similarities and differences between the two tasks, and the plug-and-play temporal feature extraction module is presented to allow the VSR-adapted model to utilize information from neighboring frames. We applied it to five representative SISR models to evaluate our method, including a generator of GAN [19], three representative SISR models [13,20,21], and a Transformer-based model [15]. Compared with state-of-the-art VSR models, our VSR-adapted models achieve competitive performance.

There are several strong points of the proposed method. Firstly, the proposed architecture of VSR-adapted models provides a novel scheme to develop VSR models. As long as a SISR model follows the common architecture, it can be easily adapted to a VSR model. It reduces the delay of applications of new SISR technologies. Secondly, with the development of VSR, better temporal feature extraction techniques will be proposed, leading to better VSR performance. It divides the development of the VSR model into two independent tasks. Thirdly, the plug-and-play characteristic enables a single model to perform both SISR and VSR tasks.

Although the VSR-adapted models show promising results, we observed some failure cases in experiments. As illustrated in Figure 11, these models fail to recover tiny details. In these cases, the contrast is low in the ground truth, and the contrast is further reduced in LR frames, making SR reconstruction very challenging. Furthermore, all VSR-adapted models fail to provide clear results.

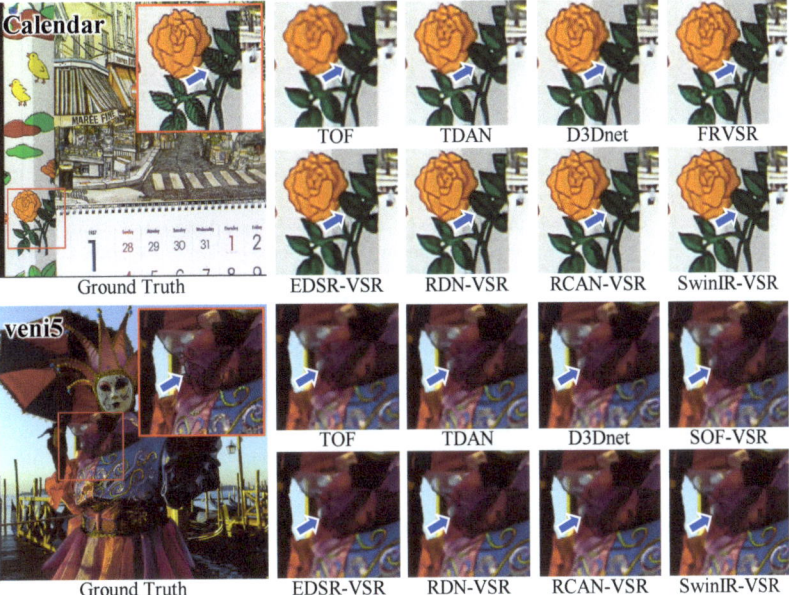

Figure 11. The Qualitative Comparison of Details in Low-Contrast Areas.

6. Conclusions

In this paper, we propose a method for adapting SISR models to the VSR task. For effectiveness on various SISR models, we summarize the common architecture of SISR models. The VSR-adapted models leverage the capability of SISR models to learn the mapping between LR and HR images. Then, the proposed plug-and-play temporal feature extraction module allows VSR-adapted models to access spatial–temporal information.

Thus, the performance in the VSR task is improved by the incorporation of the SISR model and the temporal feature extraction module. The experiments on several SISR models and benchmarks show that VSR-adapted models surpass the original SISR models. The achieved performance is positively related to the capacity of SISR models, indicating the effectiveness of the proposed method. Further, the VSR-adapted models achieved better results than the SOTA VSR models. In the future, we plan to solve the problem of poor performance in low-contrast areas.

Author Contributions: Conceptualization, W.W.; methodology, W.W.; software, W.W. and Y.H.; validation, W.W. and H.L.; formal analysis, Z.L. and R.L.; investigation, W.W., H.L. and Y.H.; resources, Z.L. and R.L.; data curation, W.W., H.L. and Y.H.; writing—original draft preparation, W.W. and H.L.; writing—review and editing, Z.L., R.L., H.L., Y.H. and W.W.; visualization, H.L., Y.H. and W.W.; supervision, Z.L.; project administration, W.W.; funding acquisition, Z.L. and R.L. All authors have read and agreed to the published version of the manuscript.

Funding: This research was funded by National Natural Science Foundation of China (61866009, 62172120, 82272075), Guangxi Science Fund for Distinguished Young Scholars (2019GXNSFFA245014), Guangxi Key Research and Development Program (AB21220037), and Innovation Project of Guangxi Graduate Education (YCBZ2022112).

Institutional Review Board Statement: Not applicable.

Informed Consent Statement: Not applicable.

Data Availability Statement: The public data used in this work are listed here: Vimeo90k http://tof-low.csail.mit.edu/ (accessed on 12 December 2022), Vid4 https://drive.google.com/file/d/1ZuvNNLgR85TV_whJoH-M7uVb-XW1y70DW/view?usp=sharing (accessed on 12 December 2022), and SPMC-11 https://pan.baidu.com/s/1PK-ZeTo8HVklHU5Pe26qUtw (accessed on 12 December 2022) (Code: 4l5r).

Conflicts of Interest: The authors declare no conflict of interest.

References

1. Yang, C.; Huang, Z.; Wang, N. QueryDet: Cascaded Sparse Query for Accelerating High-Resolution Small Object Detection. In Proceedings of the IEEE/CVF Conference on Computer Vision and Pattern Recognition, CVPR 2022, New Orleans, LA, USA, 18–24 June 2022; pp. 13658–13667. [CrossRef]
2. Shermeyer, J.; Etten, A.V. The Effects of Super-Resolution on Object Detection Performance in Satellite Imagery. In Proceedings of the IEEE Conference on Computer Vision and Pattern Recognition Workshops, CVPR Workshops 2019, Computer Vision Foundation/IEEE, Long Beach, CA, USA, 16–20 June 2019; pp. 1432–1441. [CrossRef]
3. Dong, H.; Xie, K.; Xie, A.; Wen, C.; He, J.; Zhang, W.; Yi, D.; Yang, S. Detection of Occluded Small Commodities Based on Feature Enhancement under Super-Resolution. *Sensors* **2023**, *23*, 2439. [CrossRef] [PubMed]
4. Yuan, X.; Fu, D.; Han, S. LRF-SRNet: Large-Scale Super-Resolution Network for Estimating Aircraft Pose on the Airport Surface. *Sensors* **2023**, *23*, 1248. [CrossRef] [PubMed]
5. Jumper, J.; Evans, R.; Pritzel, A.; Green, T.; Figurnov, M.; Ronneberger, O.; Tunyasuvunakool, K.; Bates, R.; Žídek, A.; Potapenko, A.; et al. Highly accurate protein structure prediction with AlphaFold. *Nature* **2021**, *596*, 583–589. [CrossRef]
6. Cheng, H.K.; Schwing, A.G. XMem: Long-Term Video Object Segmentation with an Atkinson-Shiffrin Memory Model. In Proceedings of the Computer Vision-ECCV 2022—17th European Conference, Tel Aviv, Israel, 23–27 October 2022; Proceedings, Part XXVIII; Lecture Notes in Computer Science; Avidan, S., Brostow, G.J., Cissé, M., Farinella, G.M., Hassner, T., Eds.; Springer: Berlin/Heidelberg, Germany, 2022; Volume 13688, pp. 640–658. [CrossRef]
7. Chen, Y.; Xia, R.; Zou, K.; Yang, K. FFTI: Image inpainting algorithm via features fusion and two-steps inpainting. *J. Vis. Commun. Image Represent.* **2023**, *91*, 103776. [CrossRef]
8. Imran, A.; Sulaman, M.; Yang, S.; Bukhtiar, A.; Qasim, M.; Elshahat, S.; Khan, M.S.A.; Dastgeer, G.; Zou, B.; Yousaf, M. Molecular beam epitaxy growth of high mobility InN film for high-performance broadband heterointerface photodetectors. *Surf. Interfaces* **2022**, *29*, 101772. [CrossRef]
9. Liu, H.; Ruan, Z.; Zhao, P.; Dong, C.; Shang, F.; Liu, Y.; Yang, L.; Timofte, R. Video super-resolution based on deep learning: A comprehensive survey. *Artif. Intell. Rev.* **2022**, *55*, 5981–6035. [CrossRef]
10. Haris, M.; Shakhnarovich, G.; Ukita, N. Recurrent Back-Projection Network for Video Super-Resolution. In Proceedings of the IEEE Conference on Computer Vision and Pattern Recognition, CVPR 2019, Computer Vision Foundation/IEEE, Long Beach, CA, USA, 16–20 June 2019; pp. 3897–3906. [CrossRef]

11. Haris, M.; Shakhnarovich, G.; Ukita, N. Deep Back-Projection Networks for Super-Resolution. In Proceedings of the 2018 IEEE Conference on Computer Vision and Pattern Recognition, CVPR 2018, Computer Vision Foundation/IEEE Computer Society, Salt Lake City, UT, USA, 18–22 June 2018; pp. 1664–1673. [CrossRef]
12. Tian, Y.; Zhang, Y.; Fu, Y.; Xu, C. TDAN: Temporally-Deformable Alignment Network for Video Super-Resolution. In Proceedings of the 2020 IEEE/CVF Conference on Computer Vision and Pattern Recognition, CVPR 2020, Computer Vision Foundation/IEEE, Seattle, WA, USA, 13–19 June 2020; pp. 3357–3366. [CrossRef]
13. Lim, B.; Son, S.; Kim, H.; Nah, S.; Lee, K.M. Enhanced Deep Residual Networks for Single Image Super-Resolution. In Proceedings of the 2017 IEEE Conference on Computer Vision and Pattern Recognition Workshops, CVPR Workshops 2017, IEEE Computer Society, Honolulu, HI, USA, 21–26 July 2017; pp. 1132–1140. [CrossRef]
14. Liang, J.; Fan, Y.; Xiang, X.; Ranjan, R.; Ilg, E.; Green, S.; Cao, J.; Zhang, K.; Timofte, R.; Gool, L.V. Recurrent Video Restoration Transformer with Guided Deformable Attention. *Adv. Neural Inf. Process. Syst.* **2022**, *35*, 378–393.
15. Liang, J.; Cao, J.; Sun, G.; Zhang, K.; Gool, L.V.; Timofte, R. SwinIR: Image Restoration Using Swin Transformer. In Proceedings of the IEEE/CVF International Conference on Computer Vision Workshops, ICCVW 2021, Montreal, BC, Canada, 11–17 October 2021; pp. 1833–1844. [CrossRef]
16. Xue, T.; Chen, B.; Wu, J.; Wei, D.; Freeman, W.T. Video Enhancement with Task-Oriented Flow. *Int. J. Comput. Vis.* **2019**, *127*, 1106–1125. [CrossRef]
17. Wang, L.; Guo, Y.; Liu, L.; Lin, Z.; Deng, X.; An, W. Deep Video Super-Resolution Using HR Optical Flow Estimation. *IEEE Trans. Image Process.* **2020**, *29*, 4323–4336. [CrossRef]
18. Lian, W.; Lian, W. Sliding Window Recurrent Network for Efficient Video Super-Resolution. In Proceedings of the Computer Vision-ECCV 2022 Workshops, Tel Aviv, Israel, 23–27 October 2022; Proceedings, Part II; Lecture Notes in Computer Science; Karlinsky, L., Michaeli, T., Nishino, K., Eds.; Springer: Berlin/Heidelberg, Germany, 2022; Volume 13802, pp. 591–601. [CrossRef]
19. Ledig, C.; Theis, L.; Huszar, F.; Caballero, J.; Cunningham, A.; Acosta, A.; Aitken, A.P.; Tejani, A.; Totz, J.; Wang, Z.; et al. Photo-Realistic Single Image Super-Resolution Using a Generative Adversarial Network. In Proceedings of the 2017 IEEE Conference on Computer Vision and Pattern Recognition, CVPR 2017, IEEE Computer Society, Honolulu, HI, USA, 21–26 July 2017; pp. 105–114. [CrossRef]
20. Zhang, Y.; Li, K.; Li, K.; Wang, L.; Zhong, B.; Fu, Y. Image Super-Resolution Using Very Deep Residual Channel Attention Networks. In Proceedings of the Computer Vision-ECCV 2018—15th European Conference, Munich, Germany, 8–14 September 2018; Proceedings, Part VII; Lecture Notes in Computer Science; Ferrari, V., Hebert, M., Sminchisescu, C., Weiss, Y., Eds. Springer: Berlin/Heidelberg, Germany, 2018; Volume 11211, pp. 294–310. [CrossRef]
21. Zhang, Y.; Tian, Y.; Kong, Y.; Zhong, B.; Fu, Y. Residual Dense Network for Image Super-Resolution. In Proceedings of the 2018 IEEE Conference on Computer Vision and Pattern Recognition, CVPR 2018, Computer Vision Foundation/IEEE Computer Society, Salt Lake City, UT, USA, 18–22 June 2018; pp. 2472–2481. [CrossRef]
22. Horé, A.; Ziou, D. Image Quality Metrics: PSNR vs. SSIM. In Proceedings of the 2010 20th International Conference on Pattern Recognition, Istanbul, Turkey, 23–26 August 2010; pp. 2366–2369. [CrossRef]
23. Wang, Z.; Bovik, A.; Sheikh, H.; Simoncelli, E. Image quality assessment: From error visibility to structural similarity. *IEEE Trans. Image Process.* **2004**, *13*, 600–612. [CrossRef]
24. Liu, Y.; Chu, Z.; Li, B. A Local and Non-Local Features Based Feedback Network on Super-Resolution. *Sensors* **2022**, *22*, 9604. [CrossRef]
25. Chen, Y.; Xia, R.; Yang, K.; Zou, K. MFFN: Image super-resolution via multi-level features fusion network. *Vis. Comput.* **2023**, 1–16. [CrossRef]
26. Shi, W.; Caballero, J.; Huszar, F.; Totz, J.; Aitken, A.P.; Bishop, R.; Rueckert, D.; Wang, Z. Real-Time Single Image and Video Super-Resolution Using an Efficient Sub-Pixel Convolutional Neural Network. In Proceedings of the 2016 IEEE Conference on Computer Vision and Pattern Recognition, CVPR 2016, IEEE Computer Society, Las Vegas, NV, USA, 27–30 June 2016; pp. 1874–1883. [CrossRef]
27. Lan, R.; Sun, L.; Liu, Z.; Lu, H.; Pang, C.; Luo, X. MADNet: A Fast and Lightweight Network for Single-Image Super Resolution. *IEEE Trans. Cybern.* **2021**, *51*, 1443–1453. [CrossRef] [PubMed]
28. Lan, R.; Sun, L.; Liu, Z.; Lu, H.; Su, Z.; Pang, C.; Luo, X. Cascading and Enhanced Residual Networks for Accurate Single-Image Super-Resolution. *IEEE Trans. Cybern.* **2021**, *51*, 115–125. [CrossRef] [PubMed]
29. Sun, L.; Liu, Z.; Sun, X.; Liu, L.; Lan, R.; Luo, X. Lightweight Image Super-Resolution via Weighted Multi-Scale Residual Network. *IEEE/CAA J. Autom. Sin.* **2021**, *8*, 1271–1280. [CrossRef]
30. Dong, C.; Loy, C.C.; He, K.; Tang, X. Image Super-Resolution Using Deep Convolutional Networks. *IEEE Trans. Pattern Anal. Mach. Intell.* **2016**, *38*, 295–307. [CrossRef] [PubMed]
31. He, K.; Zhang, X.; Ren, S.; Sun, J. Deep Residual Learning for Image Recognition. In Proceedings of the 2016 IEEE Conference on Computer Vision and Pattern Recognition, CVPR 2016, IEEE Computer Society, Las Vegas, NV, USA, 27–30 June 2016; pp. 770–778. [CrossRef]
32. Vaswani, A.; Shazeer, N.; Parmar, N.; Uszkoreit, J.; Jones, L.; Gomez, A.N.; Kaiser, L.; Polosukhin, I. Attention is All you Need. *Adv. Neural Inf. Process. Syst.* **2017**, *30*, 5998–6008.

33. Liu, Z.; Lin, Y.; Cao, Y.; Hu, H.; Wei, Y.; Zhang, Z.; Lin, S.; Guo, B. Swin Transformer: Hierarchical Vision Transformer using Shifted Windows. In Proceedings of the 2021 IEEE/CVF International Conference on Computer Vision, ICCV 2021, IEEE, Montreal, QC, Canada, 10–17 October 2021; pp. 9992–10002. [CrossRef]
34. Sajjadi, M.S.M.; Vemulapalli, R.; Brown, M. Frame-Recurrent Video Super-Resolution. In Proceedings of the 2018 IEEE Conference on Computer Vision and Pattern Recognition, CVPR 2018, Computer Vision Foundation/IEEE Computer Society, Salt Lake City, UT, USA, 18–22 June 2018; pp. 6626–6634. [CrossRef]
35. Sajjadi, M.S.M.; Schölkopf, B.; Hirsch, M. EnhanceNet: Single Image Super-Resolution Through Automated Texture Synthesis. In Proceedings of the IEEE International Conference on Computer Vision, ICCV 2017, IEEE Computer Society, Venice, Italy, 22–29 October 2017; pp. 4501–4510. [CrossRef]
36. Wang, X.; Chan, K.C.K.; Yu, K.; Dong, C.; Loy, C.C. EDVR: Video Restoration With Enhanced Deformable Convolutional Networks. In Proceedings of the IEEE Conference on Computer Vision and Pattern Recognition Workshops, CVPR Workshops 2019, Computer Vision Foundation/IEEE, Long Beach, CA, USA, 16–20 June 2019; pp. 1954–1963. [CrossRef]
37. Choi, Y.J.; Lee, Y.; Kim, B. Wavelet Attention Embedding Networks for Video Super-Resolution. In Proceedings of the 25th International Conference on Pattern Recognition, ICPR 2020, Milan, Italy, 10–15 January 2021; pp. 7314–7320. [CrossRef]
38. Xu, W.; Song, H.; Jin, Y.; Yan, F. Video Super-Resolution with Frame-Wise Dynamic Fusion and Self-Calibrated Deformable Alignment. *Neural Process. Lett.* **2022**, *54*, 2803–2815. [CrossRef]
39. Cao, Y.; Wang, C.; Song, C.; Tang, Y.; Li, H. Real-Time Super-Resolution System of 4K-Video Based on Deep Learning. In Proceedings of the 32nd IEEE International Conference on Application-specific Systems, Architectures and Processors, ASAP 2021, Virtual, 7–9 July 2021; pp. 69–76. [CrossRef]
40. Jo, Y.; Oh, S.W.; Kang, J.; Kim, S.J. Deep Video Super-Resolution Network Using Dynamic Upsampling Filters Without Explicit Motion Compensation. In Proceedings of the 2018 IEEE Conference on Computer Vision and Pattern Recognition, CVPR 2018, Computer Vision Foundation/IEEE Computer Society, Salt Lake City, UT, USA, 18–22 June 2018; pp. 3224–3232. [CrossRef]
41. Kim, S.Y.; Lim, J.; Na, T.; Kim, M. Video Super-Resolution Based on 3D-CNNS with Consideration of Scene Change. In Proceedings of the 2019 IEEE International Conference on Image Processing, ICIP 2019, Taipei, Taiwan, 22–25 September 2019; pp. 2831–2835. [CrossRef]
42. Isobe, T.; Li, S.; Jia, X.; Yuan, S.; Slabaugh, G.G.; Xu, C.; Li, Y.; Wang, S.; Tian, Q. Video Super-Resolution With Temporal Group Attention. In Proceedings of the 2020 IEEE/CVF Conference on Computer Vision and Pattern Recognition, CVPR 2020, Computer Vision Foundation/IEEE, Seattle, WA, USA, 13–19 June 2020; pp. 8005–8014. [CrossRef]
43. Chan, K.C.K.; Wang, X.; Yu, K.; Dong, C.; Loy, C.C. BasicVSR: The Search for Essential Components in Video Super-Resolution and Beyond. In Proceedings of the IEEE Conference on Computer Vision and Pattern Recognition, CVPR 2021, Computer Vision Foundation/IEEE, Virtual, 19–25 June 2021; pp. 4947–4956. [CrossRef]
44. Liu, Z.; Siu, W.; Chan, Y. Efficient Video Super-Resolution via Hierarchical Temporal Residual Networks. *IEEE Access* **2021**, *9*, 106049–106064. [CrossRef]
45. Lee, Y.; Cho, S.; Jun, D. Video Super-Resolution Method Using Deformable Convolution-Based Alignment Network. *Sensors* **2022**, *22*, 8476. [CrossRef]
46. Anwar, S.; Khan, S.H.; Barnes, N. A Deep Journey into Super-resolution: A Survey. *ACM Comput. Surv.* **2021**, *53*, 60:1–60:34. [CrossRef]
47. Ying, X.; Wang, L.; Wang, Y.; Sheng, W.; An, W.; Guo, Y. Deformable 3D Convolution for Video Super-Resolution. *IEEE Signal Process. Lett.* **2020**, *27*, 1500–1504. [CrossRef]
48. Liu, C.; Sun, D. On Bayesian Adaptive Video Super Resolution. *IEEE Trans. Pattern Anal. Mach. Intell.* **2014**, *36*, 346–360. [CrossRef]
49. Kingma, D.P.; Ba, J. Adam: A Method for Stochastic Optimization. In *Proceedings of the 3rd International Conference on Learning Representations, ICLR 2015, San Diego, CA, USA, 7–9 May 2015*; Conference Track Proceedings; Bengio, Y., LeCun, Y., Eds.; Springer: Berlin/Heidelberg, Germany, 2015.
50. Wen, W.; Ren, W.; Shi, Y.; Nie, Y.; Zhang, J.; Cao, X. Video Super-Resolution via a Spatio-Temporal Alignment Network. *IEEE Trans. Image Process.* **2022**, *31*, 1761–1773. [CrossRef] [PubMed]
51. Zhu, X.; Li, Z.; Lou, J.; Shen, Q. Video super-resolution based on a spatio-temporal matching network. *Pattern Recognit.* **2021**, *110*, 107619. [CrossRef]

Disclaimer/Publisher's Note: The statements, opinions and data contained in all publications are solely those of the individual author(s) and contributor(s) and not of MDPI and/or the editor(s). MDPI and/or the editor(s) disclaim responsibility for any injury to people or property resulting from any ideas, methods, instructions or products referred to in the content.

Inpainting with Separable Mask Update Convolution Network

Jun Gong [1], Senlin Luo [1], Wenxin Yu [2] and Liang Nie [2,*]

1. Information System and Security & Countermeasures Experimental Center, Beijing Institute of Technology, Beijing 100081, China
2. School of Computer Science and Technology, Southwest University of Science and Technology, Mianyang 621010, China
* Correspondence: jianglihuakai@gmail.com

Abstract: Image inpainting is an active area of research in image processing that focuses on reconstructing damaged or missing parts of an image. The advent of deep learning has greatly advanced the field of image restoration in recent years. While there are many existing methods that can produce high-quality restoration results, they often struggle when dealing with images that have large missing areas, resulting in blurry and artifact-filled outcomes. This is primarily because of the presence of invalid information in the inpainting region, which interferes with the inpainting process. To tackle this challenge, the paper proposes a novel approach called separable mask update convolution. This technique automatically learns and updates the mask, which represents the missing area, to better control the influence of invalid information within the mask area on the restoration results. Furthermore, this convolution method reduces the number of network parameters and the size of the model. The paper also introduces a regional normalization technique that collaborates with separable mask update convolution layers for improved feature extraction, thereby enhancing the quality of the restored image. Experimental results demonstrate that the proposed method performs well in restoring images with large missing areas and outperforms state-of-the-art image inpainting methods significantly in terms of image quality.

Keywords: image inpainting; image processing; separable mask update convolution; encoder-decoder network

1. Introduction

Image inpainting restores missing or damaged portions of an image, playing a crucial role in image coding and computational imaging. It fills in missing areas with plausible content, improving coding efficiency and fidelity. In computational imaging, it helps overcome challenges like occlusions and incomplete data, generating accurate scene representations. Overall, image inpainting facilitates efficient representation, transmission, and analysis of visual data, offering promising solutions for practical applications.

This technique has been widely applied in various fields, including medicine, military, and video processing, among others [1–6]. The early image inpainting methods were mainly based on traditional image processing techniques, such as texture synthesis, patch-based methods, and exemplar-based methods. Texture synthesis methods [7] typically fill in missing areas by replicating or generating textures from the surrounding image regions. Patch-based methods [8] use similar patches from the non-missing areas to fill in the missing regions. Exemplar-based methods [9], on the other hand, utilize a set of exemplar images to complete the missing regions by finding the most similar patches or structures from the exemplars. However, these early methods often suffer from a limited capability to handle complex structures and to generate realistic textures, resulting in visible artifacts and inconsistencies in the inpainted regions.

Recently, deep-learning-based approaches have emerged as the state-of-the-art for image inpainting, due to their ability to learn complex relationships and structures in the image data. These methods [10–12] typically involve using encoder-decoder networks to learn the context of the surrounding pixels, and then use this information to infer the missing content. Pathak et al. [13] were the first to apply convolutional neural networks to image restoration, and they designed a context encoder to capture the background information of images. Yang et al. [14] designed a dual-branch generator network, where one branch focuses on restoring the texture information of the image, while the other focuses on restoring the structural information, and then the results of the two branches are fused to improve the quality of the image restoration. Subsequently, due to the outstanding performance of generative adversarial networks in image restoration, many deep neural network architectures began to adopt adversarial learning strategies for image inpainting. For example, Yeh et al. [15] proposed an adversarial learning network consisting of a generator and discriminator, which can automatically generate high-quality restored images. The generator aims to generate complete images from the missing parts, while the discriminator is used to evaluate whether the generated results are similar to natural images. Iizuka et al. [16] proposed the concepts of the global discriminator and local discriminator. The global discriminator is used to detect the consistency of the overall image, while the local discriminator is used to detect the details and texture of local regions. The texture consistency of the restored results is ensured by evaluating the entire image and local regions. With the rapid development of deep learning, new technologies are constantly emerging in the field of image inpainting. For example, contextual attention mechanisms [17] can capture contextual information of different scales in the image, thereby improving the accuracy of image restoration. Partial convolution [18] only convolves known regions and ignores other missing parts, which can better handle missing areas. Gate convolution [19] can adaptively weight information from different positions to improve image restoration quality. Region normalization [20] can enhance the model's generalization ability, thus making the image restoration results more accurate and robust.

However, it is challenging to model both the texture and structure of an image using a single shared framework. To effectively restore the structure and texture information of images, researchers, such as Guo et al. [21], have proposed a novel dual-stream network called CTSDG. This approach decomposes the image inpainting task into two subtasks, namely texture synthesis and structure reconstruction, further improving the performance of image restoration. This strategy allows for better handling of different feature requirements, resulting in enhanced quality and accuracy of the restored images.

Furthermore, existing image inpainting techniques typically provide only a single restoration result. However, image inpainting is inherently an uncertain task, and its output should not be limited. To address this issue, Liu et al. [22] introduced a new approach based on the PD-GAN algorithm. They considered that the closer the hole is to the center, the higher its diversity and strength. By leveraging this idea, they achieved satisfactory restoration results. This method introduces more diversity and realism in the restoration outcomes, enabling better adaptation to different inpainting requirements.

To address the restoration of boundary and high-texture regions, Wu et al. [23] proposed an end-to-end generative model method. They first used a local binary pattern (LBP) learning network based on the U-Net architecture to predict the structural information of the missing areas. Additionally, an upgraded spatial attention mechanism was introduced as a guide and incorporated into the image inpainting network. By applying these techniques, the algorithm aims to better restore the missing pixels in boundary and high-texture regions.

The aforementioned deep learning-based inpainting methods rely on the encoder-decoder to infer the context of small missing image areas. They then infer the texture details of the missing area based on the image features of the non-missing area and use local pixel correlation to restore the damaged image area. However, when the missing area of the image becomes larger and the distance between unknown and known pixels increases, these methods can produce semantic ambiguity due to the weakening of pixel correlation. Additionally, due to the limitations of convolution kernel size and a single convolution layer, the range of extracted information is too small to capture global structural information from distant pixels. As a result, it is challenging to repair larger missing areas with more semantics directly in one step.

Mou et al. [24] proposed a novel model called a deep generalized unfolded network (DGU-Net). This model integrates gradient estimation strategies into the steps of the gradient descent algorithm (PGD) to enhance the performance of the restoration process. However, it was not successful in handling large-area missing images. This indicates that there are indeed difficulties in effectively restoring images with extensive missing regions.

Inspired by the human learning process, by first learning some simple tasks and then gradually increasing the difficulty of the task, this learning strategy, from easy to difficult, can gradually learn a better performance model. The pixels inside the region are easier to repair. Therefore, Zhang et al. [25] proposed another progressive repair method, which progresses from the border of the missing region to the center. However, the progressive repair method must update the feature map in each iteration mapping back to the RGB space, resulting in a high computational cost. In response to this problem, Li et al. [26] designed the RFR-Net model to perform progressive restoration at the image feature level. That is, the input and output of the model need to be in the same space representation, which greatly saves computational costs. However, the RFR-Net model only uses the learnable convolution kernel to perceive the edge of the damaged area, ignoring the context information outside the receptive field. There are still some problems with blurred boundaries and incorrect semantic content that lead to repair results.

Aiming at the problem of huge amount of network parameters in image inpainting, we naturally think of optimizing the network structure and reducing unnecessary network layers. This paper uses the simplified encoder-decoder as the backbone of the generator. The end-to-end one-stage network dramatically reduces the complexity of the network compared to the progressive inpainting and multi-stage networks. Nevertheless, the cost of doing this is that the network may lose some ability to capture fine-grained texture details and global structural information, especially in large missing regions. In order to improve the restoration effect, this paper proposes a separable mask update convolution to reduce the interference caused by the missing regions in the image during the restoration process.

This paper presents three main contributions in the field of image inpainting:

- Lightweight end-to-end inpainting network: The paper introduces a novel lightweight end-to-end inpainting generative adversarial network. This network architecture, consisting of an encoder, decoder, and discriminator, addresses the complexity issue present in existing inpainting methods. It enables fast and efficient image restoration while maintaining high-quality inpainting results. The streamlined network design ensures computational efficiency and practicality;
- Separable mask update convolution: The paper proposes a unique method called separable mask update convolution. By improving the specific gating mechanism, it enables automatic learning and updating of the mask distribution. This technique effectively filters out the impact of invalid information during the restoration process, leading to improved image restoration quality. Additionally, the adoption of deep separable convolution reduces the number of required parameters, significantly reducing model complexity and computational resource demands. As a result, the inpainting process becomes more efficient and feasible;
- Superior inpainting performance: Experimental results demonstrate that the proposed inpainting network surpasses existing image inpainting methods in terms of both network parameters and inpainting quality. The innovative network architecture, coupled with the separable mask update convolution, achieves superior inpainting results with fewer parameters, reducing model complexity while maintaining high-quality restorations.

2. Related Work
2.1. Attention Mechanism

The attention mechanism can help the image inpainting model to find the most similar feature block from the non-missing area of the image according to the characteristics of the missing area, thereby improving the quality of image inpainting. Yu et al. [17] added an attention mechanism to the image inpainting network. The extracted feature information is divided into foreground and background areas, and the image feature blocks are matched in a long distance according to the similarity of the foreground and background. However, this image inpainting method ignores the correlation between the internal features of the missing area of the image. Therefore, Liu et al. [27] proposed a coherent semantic attention mechanism, which effectively improves the semantic consistency of the internal features in the missing area of the image. Since the features extracted by deep and shallow layers are not the same in convolutional neural networks, Zeng et al. [28] proposed a pyramidal context encoder network. The attention transfer network can transfer the attention information obtained from the high-level semantic features to the low-level features. This model is a restoration method that acts on the feature layer, which can improve the semantic consistency of the image after restoration. Literature [26] proposes a recurrent feature reasoning network, which works on the image feature level. In the process of feature reasoning, the designed knowledge consistent attention (KCA) module is added. The attention score determines the attention score of this module in the loop process, and the current attention score is jointly determined. This method can significantly save computational costs and achieve a more refined repair result. However, the features located in the missing area usually have a significant deviation, leading to the attention module's wrong attention allocation. Finally, the model fills in incorrect texture details for some missing areas. Phutke et al. [29] applied wavelet query multi-head attention to image inpainting. Wavelet query multi-head attention is an attention mechanism that combines wavelet transforms with multi-head attention. This allows the model to attend to information from different representation subspaces at different positions, improving its ability to capture long-range dependencies and complex relationships between the input and output sequences.

2.2. Convolution Method

Convolution is a fundamental mathematical operation in deep neural networks to extract essential features from input signals or images. It has revolutionized computer vision and is widely used in various deep learning tasks, including image classification, object detection, segmentation, and image inpainting. Researchers commonly used valid convolution for feature extraction in the early stages of applying deep learning to image inpainting. During this period, they mainly focused on studying the restoration of regular square-shaped missing regions in the center of the image, as in the work of Pathak et al. [13] and Yu et al. [17]. Since the missing regions were regular, their impact on the restoration results was relatively low during the convolution kernel sliding process. However, what needs to be restored is often irregular regions. In this case, feature extraction using valid convolutions suffers interference from missing regions. Because the convolution kernel will cover many mixed windows of effective areas and invalid areas during the sliding process, this can lead to inaccurate learned features and thus affect image restoration results.

So, researchers began exploring using more advanced convolutional for image inpainting. The concept of partial convolution was first proposed by Liu et al. [18]. Partial convolution uses only valid pixels in the kernel to compute the output, ignoring invalid pixels (such as those in missing regions). This allows the convolution operation to focus on valid pixels, preventing missing regions from affecting the learned features. Partial convolution also has some limitations. One of the main limitations is that partial convolution is computationally expensive compared to regular convolution because it requires additional calculations to generate the mask. Additionally, partial convolution may not be suitable for cases where the missing regions occupy a large portion of the image because the valid pixels may not provide sufficient contextual information for restoration. Subsequently, in order to solve the problem that partial convolution cannot handle large areas of missing regions, gated convolution was proposed by Yu et al. [19]. Gated convolution is a variant of partial convolution that introduces an additional gating mechanism to control the flow of information through the convolutional kernel. The gating mechanism consists of a sigmoid function that generates a gating map to modulate the convolutional kernel's feature responses. The gating map is used to selectively pass through the valid pixels in the convolutional kernel and suppress the invalid pixels in the missing regions. Liu et al. [30] used part of the convolution kernel to process the structure and texture features of the image to generate feature information with different scales. Due to the excellent performance of gated convolution in repairing irregularly missing regions, Ma et al. [31] proposed an improved version called dense gate convolution. This method incorporates the idea of dense connections, which allows information to flow freely within the network, thereby enhancing feature propagation and utilization.

Although the above convolution method solves the problem of large-area irregular mask image competition, there is still room for improvement in image restoration quality. Moreover, the problems of parameter expansion and increased calculation consumption caused by complex convolution methods have yet to be resolved but have intensified.

2.3. Progressive Image Inpainting

In view of the weak ability of convolutional neural networks in modeling long-distance pixel correlations between known long-distance regions (background regions) and regions to be inpainted (foreground regions), progressive image inpainting methods have been widely used in recent years. Xiong et al. [32] divided the whole inpainting task into three parts in sequence: perceiving the image foreground, completing object contours, and inpainting missing regions [33]. first predicted the structural information of the missing region of the image and then repaired the image according to the predicted structural information to improve the feature structure consistency between the repaired image and the real image. An excellent residual architecture in the full-resolution residual network proposed by Guo et al. [34] is helpful for feature integration and texture prediction. Furthermore, each residual block only reconstructs the specified missing regions to ensure image quality during the progressive inpainting process. Chen et al. [35] completed the image inpainting task step by step from the perspective of pyramid multi-resolution, during which low-resolution inpainting and high-resolution inpainting are performed in a cycle. Li et al. [36] stacked the visual structure reconstruction layer in the U-Net structure containing some convolutional layers. They reconstructed the structure and visual features of the missing area in a progressive manner. In this network, the updated structural information in each visual structure reconstruction layer is used to guide the filling of feature content to gradually reduce the missing area and finally complete the restoration task. Liao et al. [37] proposed a progressive image inpainting network that uses semantic segmentation information to constrain image content. However, these progressive image inpainting methods ignore the contextual information outside the receptive field of the convolution kernel. Shi et al. [38] proposed a multi-stage progressive inpainting method that divides the inpainting process into three stages: feature extraction, interactive inpainting, and reinforcement reconstruction. They used a dual-branch structure to focus on gradually restoring texture-level features. This approach avoids the redundant computation of previous cyclic progressive inpainting methods. Liu et al. [39] also used a dual-branch structure, but instead of having high-resolution and low-resolution branches, they focused on two progressive feature extraction branches for structure and texture feature extraction. This approach allows for the maximum restoration of the image's structure and texture information.

2.4. GAN for Inpainting

Generative adversarial networks (GAN) is a deep learning model consisting of a generator and a discriminator. It has been widely used in image inpainting. The generator takes an image with missing regions as input and generates a repaired image, while the discriminator tries to distinguish between the generated repaired image and the real image. Through adversarial training, the generator gradually learns to generate realistic repaired images that visually resemble the real images. The application of GAN in image inpainting has the advantage of generating natural and preserving structural and texture features in the repaired results. In recent years, researchers have proposed different GAN-based image inpainting methods.

Guo et al. [21] proposed a novel two-stream network that models structure-constrained texture synthesis and texture-guided structure reconstruction in a coupled manner. The method, named "Conditional Texture and Structure Dual Generation (CTSDG)," incorporates a bi-directional gated feature fusion (Bi-GFF) module to exchange and combine structure and texture information, and a contextual feature aggregation (CFA) module to refine the generated contents using region affinity learning and multi-scale feature aggregation. Li et al. [40] introduced a novel multi-level interactive Siamese filtering (MISF) technique that combines image-level predictive filtering and semantic filtering on deep feature levels. Their method contains two branches: a kernel prediction branch (KPB) and a semantic and image filtering branch (SIFB). These branches interactively exchange information, with SIFB providing multi-level features for KPB and KPB predicting dynamic kernels for SIFB. This method leverages effective semantic and image-level filling for high-fidelity inpainting and enhances generalization across scenes. Chen et al. [3] proposed a feature fusion and two-step inpainting approach (FFTI). The method utilizes dynamic memory networks (DMN+) to fuse external and internal features of the incomplete image and generate an incomplete image optimization map. A generation countermeasure generative network with gradient penalty constraints is constructed to guide the rough repair of the optimized incomplete image and obtain a rough repair map. Finally, the coarse repair graph is optimized using the coherence of relevant features to obtain the final fine repair graph. Xia et al. [41] proposed an effective image inpainting method called a repair network and optimization network (RNON), which utilizes two mutually independent generative adversarial networks (GANs). The image repair network module focuses on repairing irregular missing areas using a generator based on a partial convolutional network, while the image optimization network module aims to solve local chromatic aberration using a generator based on deep residual networks. The synergy between these two network modules improves the visual effect and image quality of the inpainted images.

Although these methods have made significant progress in image texture and structure restoration, they also have certain limitations. The multi-stage, multi-branch, and multi-network nature of these methods leads to increased computational resource consumption and longer computation time. CTSDG, while advantageous in coupled texture synthesis and structure reconstruction, may have limitations when dealing with large-scale corruptions or missing regions spanning important areas of the image. The computational complexity of the Bi-GFF and CFA modules may restrict real-time performance in certain applications. MISF, which focuses on semantic filtering rather than fine-grained texture reconstruction, may experience a sharp performance decline when dealing with large missing areas requiring detailed texture recovery. The effectiveness of FFTI may depend on the quality and completeness of the external and internal features used for fusion, making it susceptible to interference from irrelevant information. RNON, which utilizes two independent GANs, requires more computational resources and time for training and faces an increased risk of mode collapse. The method's repaired results also tend to exhibit over-smoothing.

3. Approach

Like a person completing a jigsaw puzzle alone, the image inpainting algorithm fills in the missing area by piecing together the surrounding pixels and keeping the contextual semantic information consistent with the image structure during the filling process. However, in the related work mentioned above, although progressive inpainting uses surrounding pixels to progressively restore missing pixels, they cannot maintain contextual image semantics and structural information well. At the same time, these algorithms have a large number of parameters and are computationally expensive. Therefore, this paper proposes a U-Net-like codec network with separable mask update convolution, significantly reducing network complexity.

3.1. Network Structure

The proposed SMUC-net in this paper is a deep learning-based image restoration model whose backbone is a codec serving as the generator. Including components such as a discriminator, encoder, and decoder constructs an end-to-end learning framework that allows the entire restoration process to be completed within a unified framework. The overall structure of the SMUC-net is shown in Figure 1. Specifically, the encoder adopts separable mask update convolution modules and region normalization modules, which can effectively extract image feature information and optimize computation efficiency. Next, the image feature vector undergoes processing through eight residual blocks, which can effectively increase the depth and flexibility of the model while avoiding the problem of gradient vanishing. Finally, the decoder consists of separable mask update convolution layers and region normalization modules and uses the tanh function to activate the output result, obtaining the restored image.

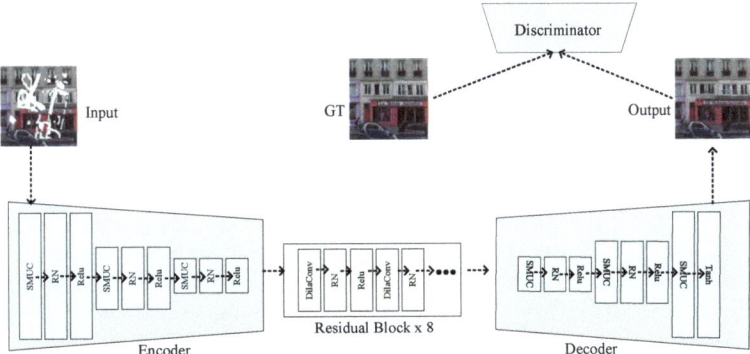

Figure 1. The structure of SMUC-net.

In SMUC-net, the discriminator evaluates the similarity between the generated restored and original images, providing feedback mechanisms for model optimization. Among them, the loss function is the core of model optimization. This paper adopts multiple loss functions such as L1 loss, adversarial loss, perceptual loss, and style loss to train the model(the loss function is consistent with [20]). The L1 loss measures the pixel-level distance between the restored and original images, while the adversarial loss encourages the generator to produce more realistic restored images. The perceptual loss ensures the perceptual quality of the restored image by comparing the feature representations of the restored image and the original image. The style loss can make the restored image better conform to the style characteristics.

In recent years, deep learning has dramatically advanced the field of image restoration due to its remarkable performance in various applications. However, the challenges remain since the image restoration problem is inherently complex and challenging. SMUC-net represents a significant step forward in addressing these challenges, providing a powerful tool for restoring damaged or missing image information. Moreover, combining various loss functions and using a unified end-to-end learning framework in SMUC-net further enhance its effectiveness and versatility. Future research can explore improving the efficiency and accuracy of image restoration models like SMUC-net, extending their applicability to even more complex scenarios.

3.2. Separable Mask Update Convolution Modules

For irregular image inpainting tasks, missing regions may have arbitrary shapes and sizes, which means that traditional vanilla convolution-based inpainting algorithms are often incompetent. The following is the operation formula of vanilla convolution:

$$(f * g)(x, y) = \sum_{i=-k}^{k} \sum_{j=-k}^{k} f(x - i, y - j) \cdot g(i, j) \tag{1}$$

Among them, f is the input image. g is the filter. k is the radius of the filter. (x, y) is the pixel position of the output image. The indices i and j represent the spatial location within the kernel matrix.

This formula says that each pixel value in the output image is the weighted sum of the filter at that location and surrounding pixels. It is not difficult to see from the formula that for the coordinate point position (x, y) of each channel of the input image f, the ordinary convolution will use the same shape of the filter g to perform convolution operations on it. This is because in ordinary convolution, the parameters of the filter are fixed independent of the pixel values in the input image. Therefore, no matter what the pixel values in the input image are, the same filter will be used for convolution. However, in tasks such as irregular image repair, since the missing area may have arbitrary shape and size, while the filter used by ordinary convolution is fixed, it is difficult to adapt to missing areas of different shapes and sizes, resulting in poor repair results.

In order to solve the above problems, it is necessary to control the interference of invalid information in the missing area of the image on the convolution result. Therefore, Liu et al. [18] proposed the concept of partial convolution. The calculation process of partial convolution needs a binary mask to assist. This mask consists only of 0 and 1. The mask position corresponding to the position where the pixel is 0 in the input image is also 0, and the corresponding mask position is 1 in other cases. Partial convolution uses a mask map to mark the areas of the input image that contain missing pixels. During convolution, only the valid pixel-containing areas are used for the convolution operation, while the invalid areas are excluded. This ensures that the missing areas do not interfere with the convolution result. Furthermore, a new mask image is generated during partial convolution, which guides the convolution operation of the subsequent layers. The operation flow of partial convolution is shown on the left side of Figure 2. Firstly, partial convolution is used to perform convolutional calculations on the input image, resulting in a new feature map. During this process, only the areas containing valid pixels are involved in the calculation, while the missing areas are excluded to avoid interfering with the convolution result. Then, the updated mask is used to perform a dot product with the feature map to obtain the input of the next layer in the network. Meanwhile, the updated mask is also used as the input for the next layer to guide the subsequent convolution operation. The partial convolution operation formula and mask update rules can be expressed as follows:

$$x' = \begin{cases} \sum\sum w \cdot \left(x \odot \frac{1}{\text{sum}(m)} \right), & \text{if sum}(m) > 0 \\ 0, & \text{otherwise} \end{cases} \tag{2}$$

w represents the weight of the current convolution filter. x is the characteristic value of the current sliding window. $\odot$ denotes element-wise multiplication. m represents the binary mask used for marking. (x, y) is the pixel position of the mask. After each partial convolution, mask-update follows the following strategy:

$$\hat{m}(x,y) = \begin{cases} 1 & \text{if sum}(m) > 0 \\ 0 & \text{otherwise} \end{cases} \quad (3)$$

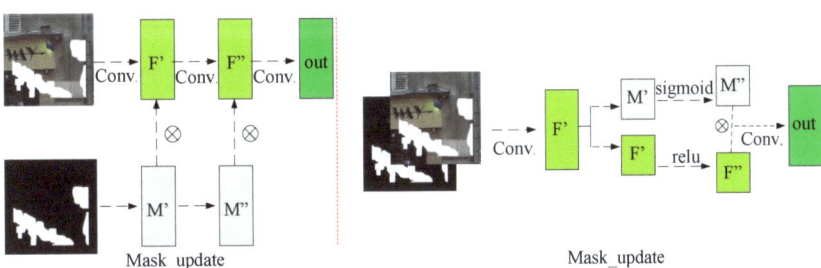

Figure 2. Illustration of partial convolution (**left**) and gate convolution (**right**).

The process of updating the mask is illustrated in Figure 3. When the mask region covered by the current convolutional window contains valid pixels, the corresponding region in the updated mask is marked as 1. However, the current masking method has a problem of considering all pixels as either valid or invalid, without considering the number of valid pixels. Therefore, in some cases, regions with only one valid pixel and regions with nine valid pixels are considered to have the same value. This is obviously unreasonable, especially as the network gets deeper, the actual information carried by regions of valid pixels can be very limited, as shown in the green region in the figure. The red area contains more effective pixels.

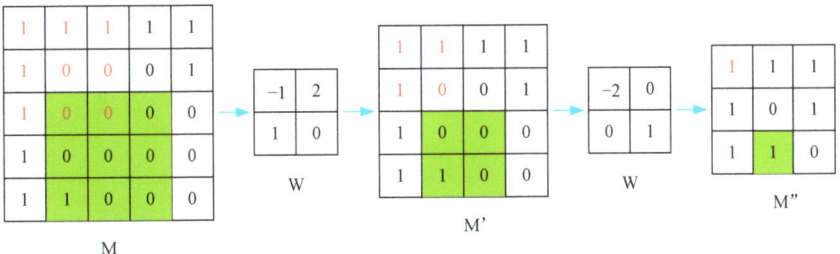

Figure 3. An illustration is provided to demonstrate the mask update mechanism employed by the partial convolution method.

Based on the concept mentioned above, Yu et al. [19] proposed a novel approach to update the mask in the feature maps. They classified the feature maps into two groups based on the number of channels. They applied either sigmoid activation or relu activation on each group, as depicted on the right side of Figure 2. Specifically, the sigmoid activation operation was named GATE. It generated a weight map, also known as a 'soft mask', through which the pixels with lower weight values were deemed to have a higher probability of containing invalid information. The soft mask was then multiplied with the activated feature map to reduce the confidence of invalid pixels in the feature map, thereby allowing the network to focus on the most informative regions. This process improved the quality of feature representations and helped eliminate the negative influence of noise and irrelevant information.

The gate convolution proposed by Yu et al. [19] is undoubtedly effective in improving the quality of feature representations. However, this approach also presents two notable challenges that need to be addressed. Firstly, the feature maps are split into two groups, which require twice as many convolution kernels and double the number of parameters in the model. This operation increases the computational cost and storage requirements, making deploying the model on resource-limited devices challenging. Secondly, the equal division of feature maps may reduce the feature space available for learning and limit the expressive power of the model.

This paper proposed a novel approach called separable mask update convolution to overcome these limitations. This approach addresses the challenges mentioned above by introducing a two-step process separating the convolutional and gating operations. Specifically, the separable mask update convolution first applies a regular convolutional operation to the input feature map, generating a set of intermediate feature maps. Then, a gating operation is performed on the intermediate feature maps to obtain a set of weight maps. By separating the two operations, the separable mask update convolution can reduce the number of convolution kernels and parameters in the model while achieving similar or even better performance than the original gate convolution approach. Moreover, the separable mask update convolution approach allows more flexibility in designing the model architecture and improves the model's ability to learn complex representations.

Based on the operation principle shown in Figure 4, the separable mask update convolution method follows a few steps. Firstly, the group parameter of the convolution kernels is set to be the same as the number of input channels, which results in an equal number of output feature maps as the input channels. Let us assume the number of input channels is N_c. Therefore, N_c feature maps are generated.

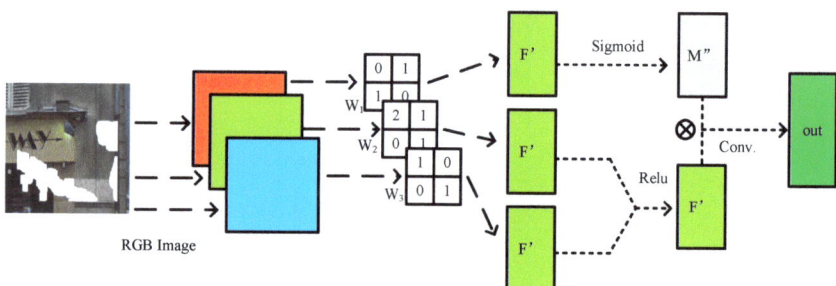

Figure 4. Illustration of our separable mask update convolution.

Next, the N_c feature maps are divided into two groups with a proportion of $N_c - 1 : 1$. The relu function activates the first group. In contrast, the sigmoid function activates the second group. The reason for using two different activation functions is to provide diverse nonlinear transformations to the feature maps. After activation, the two groups are multiplied to obtain the weighted feature maps.

Finally, the weighted feature maps are passed through the convolutional layer, which consists of filters with a kernel size equal to 1. This convolutional layer helps to expand the output channels and generates a new set of feature maps with increased depth. By using the separable mask update convolution method, the model can learn more complex representations with fewer parameters, resulting in better performance and faster convergence during training.

The convolution method proposed in this paper can significantly reduce the number of parameters compared to gate convolution. It also optimizes the number of feature maps required for mask update and improves the information utilization rate. Below are the formulas for calculating parameters in gated convolution and separable mask update convolution.

The calculation of parameters for gate convolution [19] can be expressed as:

$$\begin{aligned} N_{gc} &= K_{size} \times K_{size} \times N_{out} \times N_{in} \\ &= (K_{size} \times K_{size} \times N_{in}) \times N_{out} \end{aligned} \qquad (4)$$

The calculation of parameters for separable mask update convolution can be expressed as:

$$\begin{aligned} N_{smuc} &= K_{size} \times K_{size} \times N_{in} + N_{out} \times 1 \times 1 \times N_{in} \\ &= (K_{size} \times K_{size} + N_{out}) \times N_{in} \\ &= (K_{size} \times K_{size} \times N_{in}) \times (1 + \frac{N_{out}}{K_{size} \times K_{size}}) \end{aligned} \qquad (5)$$

The formulas show that the number of parameters required for the two convolution methods, gate convolution (GC) and separable mask update convolution (SMUC). The variables N_{gc} and N_{smuc} respectively represent the number of parameters required for each method. The kernel size is denoted by K_{size}, while N_{in} and N_{out} represent the number of input and output channels. It can be observed that when $K_{size} = 1$, SMUC has more parameters than GC due to the presence of an additional N_{in} term. However, for larger kernel sizes, SMUC requires fewer parameters than GC.

3.3. Region Normalization

The separable mask update convolution effectively reduces the impact of invalid information in the missing region on the restoration results during the convolution process. In the normalization layer, regular normalization methods cannot completely avoid the interference of invalid information in the missing region on the restoration results, especially when the missing region is large. Traditional normalization methods usually normalize the pixel values of each feature map to reduce the covariance between feature maps, thereby enhancing the robustness and generalization ability of the network. However, for the missing region, as the pixel values in the missing region are usually zero or very small, such normalization methods often cannot effectively reduce the invalid information in the missing region but may increase the impact of noise, further affecting the quality of the restoration results. To solve this problem, some unique normalization methods for the missing region have emerged in recent years, such as the normalization method based on local variance and the normalization method based on masks. These methods can better remove invalid information in the missing region through special processing of the missing region, thereby improving the quality of the restoration results.

In this paper, a novel technique called Region Normalization (RN) [20] is introduced to address the challenge of mean and variance shifts in the normalization process. The method is specifically designed for use in the early layers of the inpainting network, where the input feature contains large corrupted regions that result in significant mean and variance shifts. RN addresses this issue by separately normalizing and transforming the corrupted and uncorrupted regions, thus preventing information mixing. In specific operations, the region normalization method divides each input feature map according to its four dimensions (N, C, H, W). Then it divides it into damaged and undamaged regions based on whether there is damaged data in each region. As shown in Figure 5, the height and width of each feature map in the batch can be divided into multiple block regions. The green area represents the damaged data, and the blue area represents the undamaged data, which are normalized separately.

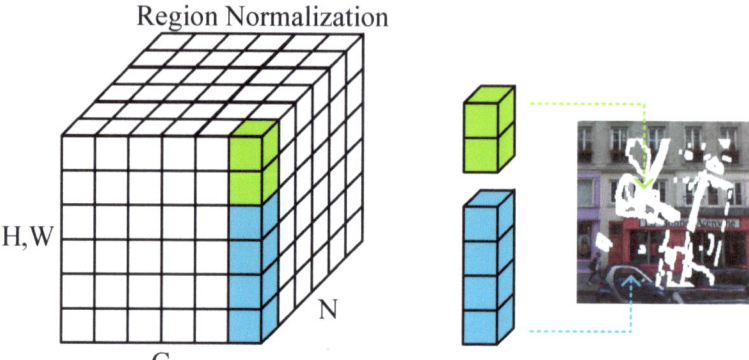

Figure 5. This figure demonstrates the proposed Region Normalization (RN) method, which applies different normalization parameters to pixels of different regions. Specifically, pixels within the same region, represented by green or blue color, share the same mean and variance during normalization. Notably, the corrupted and uncorrupted regions of the input image are normalized by different mean and variance parameters.

3.4. Loss Function

When performing image restoration, the loss function is a crucial part. By defining an appropriate loss function, we can guide the model to learn how to better restore the image. In this paper, various loss functions are adopted for image inpainting, including perceptual loss, reconstruction loss, adversarial loss and style loss. These four loss functions correspond to the reconstruction error, feature similarity, adversariality, and style similarity between the inpainted image and the real image, respectively. The total loss of the generator can be expressed as a weighted sum of these four loss functions, where each loss function has its own weight. The following are the specific formulas of reconstruction loss, perceptual loss, confrontation loss and style loss:

$$\mathcal{L}_{recon} = \|\hat{y} - y\|_2^2 \tag{6}$$

$$\mathcal{L}_{perceptual} = \frac{1}{N}\sum_{i=1}^{N}\|\phi_i(\hat{y}) - \phi_i(y)\|_2^2 \tag{7}$$

$$\mathcal{L}_{adv}(G,D) = \log D(y) + \log(1 - D(G(z))) \tag{8}$$

$$\mathcal{L}_{style} = \frac{1}{N}\sum_{i=1}^{N}\|\text{Gram}(\phi_i(\hat{y})) - \text{Gram}(\phi_i(y))\|_2^2 \tag{9}$$

Among them, $\hat{y}$ is the image generated by the generator, y is the real image, ϕ_i represents the feature representation of the i-th layer in the pre-training model, specifically pool1, pool2, and pool3 layers. Therefore, in this paper, $L = 3$, N_i represents the number of elements in ϕ_i, D is the discriminator, G is the generator, z is the input image, and Gram represents the Gram matrix. The total loss is then expressed as:

$$\mathcal{L}_{adv} = \frac{1}{N}\sum_{i=1}^{N}\log(D(\hat{y})) \tag{10}$$

The weight coefficients $\alpha, \beta, \gamma, \delta$ are used to control the contribution of each loss function to the overall loss. The total loss function includes four losses, which constrain the generator from different perspectives and effectively improve the quality and effectiveness of image restoration. In this paper, α, β, γ, and δ are set to 1, 0.1, 0.1, and 250, respectively.

4. Experiment

4.1. Experiment Setup

For our experiments, we chose to use three common datasets for image inpainting, namely Paris Street View [13], Celeba-HQ [42] and Places2 [43]. The Paris Street View dataset, proposed by researchers from ParisTech and INRIA (French National Institute for Research in Computer Science and Automation), consists of 15,000 high-resolution images capturing street views and buildings in Paris. This dataset is commonly used in research on scene understanding, image inpainting, and image synthesis in computer vision and image processing tasks.

The Celeba-HQ dataset, introduced by Ziwei Liu et al. in 2018, is an extension of the Celeba (Celebrities Attributes) dataset with higher quality images. It contains 30,000 high-quality images of celebrity faces that have been carefully curated and processed. The Celeba-HQ dataset is widely used for training and evaluating computer vision algorithms related to face recognition, face generation, and face inpainting tasks.

The Places2 dataset is a widely used large-scale image dataset proposed by researchers from the International Computer Science Institute (ICSI) and the Berkeley Vision and Learning Center (BVLC). It comprises over one million carefully curated images, capturing diverse real-world scenes such as indoor and outdoor environments, natural landscapes, urban street views, and office spaces. These high-quality images, with resolutions of 256×256 or 512×512 pixels, exhibit rich semantic and visual diversity, covering various scene types, lighting conditions, and perspectives. The Places2 dataset serves as a vital benchmark for scene understanding, image generation, image classification, and other computer vision tasks, enabling researchers to train and evaluate their models and driving advancements in the field.

In the study, the first 14,900 images from the Paris Street View dataset were used for training the model, while 100 images were reserved for testing. As for the Celeba-HQ dataset, the first 28,000 images were used for training, and the remaining 2000 images were used for testing. This data split ensures that the model is exposed to representative image samples during both training and testing, enabling accurate assessment of the performance and generalization capability of the image inpainting algorithm across different datasets. For the Places2 dataset, we followed the official partitioning of training and testing sets. We created our own training set by selecting the first 100,000 images from the complete Places2 training set. Similarly, our testing set was formed by selecting the first 2000 images from the complete Places2 testing set.

To create the masks for our experiments, we used the irregular mask dataset [18], which consists of a variety of randomly generated masks with different shapes and sizes. Liu et al. created and released the dataset of irregular masks when they proposed partial convolution. It has become one of the most widely used public datasets for irregular mask image inpainting among existing image restoration methods. We divided the masks into five categories based on the proportion of missing area, namely 10–20%, 20–30%, 30–40%, 40–50%, and 50–60%. Each category contained 2000 masks.

To train our model, we used a single NVIDIA GeForce GTX 1080Ti graphics card and set the number of epochs to 10. We continued training until the model converged and achieved satisfactory results on our test dataset.

4.2. Quantitative Comparison

In addition to describing the proposed method, this paper also compared it with other commonly used image restoration methods, including Region Normalization (RN) [20], Conditional Texture and Structure Dual Generation (CTSDG) [21], and Multi-level Interactive Siamese Filtering (MISF) [40]. These comparison methods have shown good performance in recent years in the field of image restoration. The comparison was carried out on two datasets, Celeba-HQ and Paris Street View, and the test results were presented in Tables 1 and 2, respectively. The two metrics used to evaluate the performance were

PSNR and SSIM, which reflect the pixel similarity and structural similarity between the inpainting results and the original image, respectively.

Table 1. This is a demonstration of the quantitative results of the method in this paper at different defect scales on the CelebA-HQ dataset (M = 2^{20}).

	Mask	RN [20]	CTSDG [21]	MISF [40]	Ours
PSNR ↑	10–20%	29.339	29.842	29.868	**31.472**
	20–30%	26.344	26.550	27.154	**28.321**
	30–40%	24.060	24.652	24.993	**26.053**
	40–50%	22.072	23.122	23.185	**24.420**
	50–60%	20.274	20.459	20.455	**21.578**
SSIM ↑	10–20%	0.919	0.935	0.933	**0.959**
	20–30%	0.866	0.878	0.889	**0.926**
	30–40%	0.811	0.832	0.838	**0.889**
	40–50%	0.749	0.778	0.780	**0.876**
	50–60%	0.667	0.686	0.687	**0.814**
Parameters ↓		25.34 M	31.26 M	27.60 M	**11.02 M**

Table 2. This is a demonstration of the quantitative results of the method in this paper at different defect scales on the Paris street view dataset (M = 2^{20}).

	Mask	RN [20]	CTSDG [21]	MISF [40]	Ours
PSNR ↑	10–20%	29.237	30.375	30.042	**31.732**
	20–30%	26.678	27.188	27.465	**28.680**
	30–40%	24.517	25.424	26.059	**26.934**
	40–50%	22.556	23.412	24.057	**25.192**
	50–60%	20.424	20.844	21.416	**22.243**
SSIM ↑	10–20%	0.912	0.930	0.926	**0.945**
	20–30%	0.848	0.875	0.877	**0.928**
	30–40%	0.781	0.819	0.833	**0.873**
	40–50%	0.707	0.743	0.761	**0.832**
	50–60%	0.598	0.647	0.655	**0.792**
Parameters ↓		25.34 M	31.26 M	27.60 M	**11.02 M**

Finally, we compared the multi-level interactive Siamese filtering (MISF) [40], repair network and optimization network (RNON) [41], and features fusion and two-steps inpainting (FFTI) [3] methods on the comprehensive Places2 dataset. In this comparative experiment, we excluded the experiments on the extreme conditions of extremely small and extremely large missing areas. The experimental results were focused on the common range of 20% to 50% missing regions. The experimental results are presented in Table 3.

The results show that the proposed method outperformed the comparison methods in terms of PSNR and SSIM on both datasets, especially in the 10% to 60% range of missing area. For example, on Celeba-HQ, the proposed method achieved a PSNR improvement of 1.06–1.6 dB and an SSIM improvement of 0.026–0.127, depending on the scale of the missing area. On Paris Street View, the PSNR improvement was 0.827–1.69 dB and the SSIM improvement was 0.019–0.137. These results indicate that the proposed method can recover more structural information, especially when repairing large missing areas.

On the comprehensive Places2 dataset, our proposed method continues to exhibit a distinct advantage over recent state-of-the-art approaches in the restoration of large-scale missing image regions. Particularly for missing areas exceeding 30%, as the extent of the missing region increases, our method consistently outperforms other techniques in terms of both PSNR and SSIM metrics. Furthermore, in the restoration of small-scale missing regions (below 30%), although our method may not surpass the performance of the highly effective FFIT method, the discrepancy in PSNR values between the two approaches remains minimal.

Table 3. This is a demonstration of the quantitative results of the method in this paper at different defect scales on the Places2 dataset (M = 2^{20}).

	Mask	MISF [40]	RNON [41]	FFTI [3]	Ours
PSNR ↑	20–30%	26.115	26.742	**27.657**	27.510
	30–40%	24.260	25.944	26.387	**26.957**
	40–50%	22.140	23.386	23.671	**24.327**
SSIM ↑	20–30%	0.863	0.894	**0.909**	0.896
	30–40%	0.795	0.846	0.862	**0.883**
	40–50%	0.741	0.798	0.811	**0.872**
Parameters ↓		27.60 M	31.46 M	34.20 M	**11.02 M**

Furthermore, the proposed method has fewer parameters compared to the RN method, which is a large-scale network with the fewest parameters. Additionally, when compared to the RNON method, our method only has one-third of the parameter count. In comparison to the FFIF method, the proposed approach reduces the number of parameters by nearly two-thirds. Hence, the proposed method not only demonstrates better performance but also exhibits a more compact structure.

4.3. Qualitative Comparison

Figures 6 and 7 show the restoration results of two datasets under different missing ratios. The results demonstrate that our proposed method outperforms other methods in restoration effectiveness at any scale. It is worth noting that the performance of the existing methods and our proposed method vary when dealing with different types of images. For example, on the CelebA-HQ dataset, which consists of facial images, the SMUC-net produces more natural-looking results with softer facial contours. In contrast, the other methods tend to generate slightly more artificial-looking images. This indicates that our proposed method is more effective in preserving the natural features of facial images.

(a) Input (b) RN (c) CTSDG (d) MISF (e) Ours (f) GT

Figure 6. The result display of our method on CelebA-HQ dataset under different missing data scales.

(a) Input (b) RN (c) CTSDG (d) MISF (e) Ours (f) GT

Figure 7. This is the result display of our method on Paris street view dataset under different missing data scales.

In contrast, on the Paris Street View dataset, which includes a variety of urban scenes, our method performs better in restoring texture details of objects such as branches and windows. This is due to our method's ability to recover the structural information of the missing areas, which is crucial in restoring the texture details of objects.

Another significant advantage of our method is its ability to handle large missing areas. As demonstrated in Figure 7, our method can effectively recover the structural information of the missing areas and generate more realistic images than other methods, even when up to 60% of the image is missing.

As shown in Figure 8, on the Places2 dataset, it is easy to observe that our proposed restoration method outperforms other methods in terms of preserving more detailed information in large-scale irregular missing regions. Notably, it effectively restores finer details such as the texture on buildings or the intricate details of a child riding a toy horse.

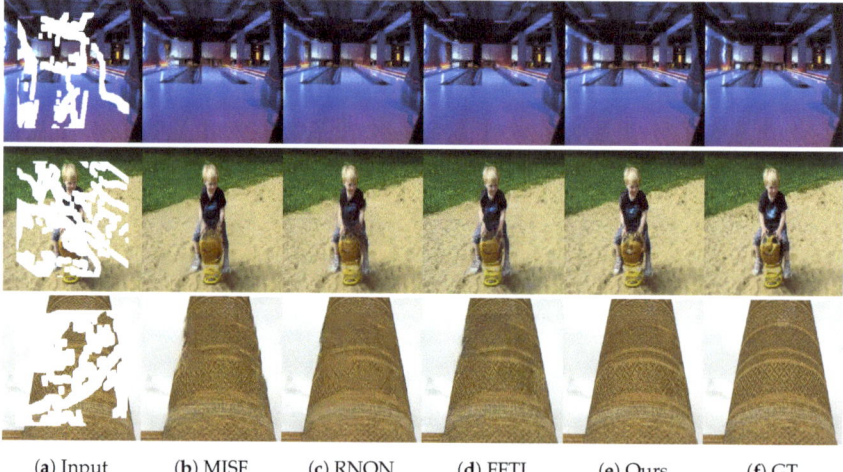

(a) Input (b) MISF (c) RNON (d) FFTI (e) Ours (f) GT

Figure 8. This is the result display of our method on Places2 dataset under different missing data scales.

However, it should be noted that our method may still encounter challenges in restoring certain types of objects, such as text or other highly structured elements. In such cases, our method may be unable to repair them successfully. Nevertheless, our method produces fewer artifacts and more realistic results compared to other methods.

In summary, our proposed method outperforms existing methods in restoring natural features and texture details, handling large missing areas, and producing more realistic results with fewer artifacts. These findings demonstrate the potential of our method for various applications, such as image editing and restoration.

4.4. Ablation Study

To validate the effectiveness of the proposed method in this paper, we conducted ablation experiments on the CelebA dataset. The mask used in this case is based on a general damage range of 30% to 40%. Following the proposed repair method, we restored separable mask update convolutions to ordinary convolutions and normalized the region normalization layer to a standard normalization layer as the base model. Then, we separately added separable mask update convolutions and region normalization layer for training and finally added both methods to the base model for training. The final model we obtained is the proposed repair method in this paper. The experimental results are shown in Table 4.

Table 4. This table provides a detailed display of the inpainting results of different network models. The base model does not include the separable mask update convolution and region normalization methods, while the SMUC-net includes both mechanisms (M = 2^{20}).

Models	PSNR	SSIM	Parameters
Base Model	23.76	0.799	12.83 M
Base Model+SGC	25.67	0.855	**11.02 M**
Base Model+RN	24.94	0.832	12.83 M
SMUC-net	**25.92**	**0.871**	**11.02 M**

After conducting experimental comparisons, we see that both replacing separable mask update convolution and region normalization layers can significantly improve the restoration performance compared to the original basic network architecture. This indicates that both methods are effective in improving restoration performance. Because both separable mask update convolutions and region normalization can reduce the interference of invalid pixels in the damaged area to some extent, they contribute to improving the repair results. It is worth noting that the core separable mask update convolution module proposed in this paper also plays an essential role in reducing network parameters. Ultimately, our experimental results show that the proposed SMUC-net achieves the best results in both restoration performance and network parameter count.

5. Conclusions

This article proposes a simple encoder-decoder network that combines separable mask update convolutions and region normalization techniques to improve image restoration. The network parameters are significantly reduced using separable mask update convolutions instead of traditional convolution operations. Additionally, the separable mask update mechanism can preserve more feature information and reduce the impact of invalid pixels by providing different weights to masked and unmasked areas, further enhancing the restoration effect.

Furthermore, the article introduces the region normalization technique to provide different means and variances for masked and unmasked areas. This method can reduce the influence of masked areas on the restoration results, thereby improving the accuracy of image restoration. Through experimental comparisons, we found that the proposed method achieved a good restoration effect and network parameter quantity results.

Experimental results on the Celeba-HQ and Paris Street View datasets show that our proposed method outperforms FFTI by 1.06–1.6 dB and 0.827–1.69 dB in terms of PSNR and by 2.6% to 12.7% and 1.9% to 13.7% in terms of SSIM under damage rates of 10% to 60%. Moreover, our method successfully reduces the parameter quantity by 16.58 M, making it the model with the minor parameters but the best restoration results.

The image inpainting method proposed in this paper has achieved significant improvements in terms of network parameters and inpainting quality. However, the main limitation of our approach is it lacks interactivity. A possible future direction could be to incorporate user guidance information into the inpainting process, which may provide more opportunities for user participation and customization. In addition, robot painting [44,45] is also a promising application direction. In practical applications, our image inpainting method can assist robots in better filling in missing image content.

Author Contributions: Conceptualization, J.G. and S.L.; methodology, J.G.; validation, J.G., S.L. and W.Y.; formal analysis, J.G.; investigation, J.G.; data curation, W.Y.; writing—original draft preparation, J.G.; writing—review and editing, J.G.; visualization, J.G.; supervision, L.N.; project administration, L.N. All authors have read and agreed to the published version of the manuscript.

Funding: This research received no external funding.

Institutional Review Board Statement: Not applicable.

Informed Consent Statement: Not applicable.

Data Availability Statement: Not applicable.

Conflicts of Interest: The authors declare no conflict of interest.

Abbreviations

The following abbreviations are used in this manuscript:

SMUC	Separable Mask Update Convolution
GC	Gate Convolution
RN	Region Normalization
CTSDG	Conditional Texture and Structure Dual Generation
MISF	Multi-level Interactive Siamese Filtering
RNON	Repair Network and Optimization Network
FFTI	Features Fusion and Two-steps Inpainting

References

1. Zhang, Y.; Zhao, P.; Ma, Y.; Fan, X. Multi-focus image fusion with joint guided image filtering. *Signal Process. Image Commun.* **2021**, *92*, 116128. [CrossRef]
2. Wang, N.; Zhang, Y.; Zhang, L. Dynamic selection network for image inpainting. *IEEE Trans. Image Process.* **2021**, *30*, 1784–1798. [CrossRef]
3. Chen, Y.; Xia, R.; Zou, K.; Yang, K. FFTI: Image inpainting algorithm via features fusion and two-steps inpainting. *J. Vis. Commun. Image Represent.* **2023**, *91*, 103776. [CrossRef]
4. Liu, K.; Li, J.; Hussain Bukhari, S.S. Overview of image inpainting and forensic technology. *Secur. Commun. Networks* **2022**, *2022*, 9291971. [CrossRef]
5. Phutke, S.S.; Murala, S. Image inpainting via spatial projections. *Pattern Recognit.* **2023**, *133*, 109040. [CrossRef]
6. Zhang, L.; Zou, Y.; Yousuf, M.H.; Wang, W.; Jin, Z.; Su, Y.; Seokhoon, K. BDSS: Blockchain-based Data Sharing Scheme with Fine-grained Access Control and Permission Revocation in Medical Environment. *Ksii Trans. Internet Inf. Syst.* **2022**, *16*, 1634–1652.
7. Bugeau, A.; Bertalmio, M. Combining Texture Synthesis and Diffusion for Image Inpainting. In Proceedings of the VISAPP 2009-Proceedings of the Fourth International Conference on Computer Vision Theory and Applications, Lisboa, Portugal, 5–8 February 2009; pp. 26–33.
8. Ružić, T.; Pižurica, A. Context-aware patch-based image inpainting using Markov random field modeling. *IEEE Trans. Image Process.* **2014**, *24*, 444–456. [CrossRef]
9. Liu, Y.; Caselles, V. Exemplar-based image inpainting using multiscale graph cuts. *IEEE Trans. Image Process.* **2012**, *22*, 1699–1711.
10. Huang, L.; Huang, Y. DRGAN: A dual resolution guided low-resolution image inpainting. *Knowl. Based Syst.* **2023**, *264*, 110346. [CrossRef]
11. Ran, C.; Li, X.; Yang, F. Multi-Step Structure Image Inpainting Model with Attention Mechanism. *Sensors* **2023**, *23*, 2316. [CrossRef]

12. Li, A.; Zhao, L.; Zuo, Z.; Wang, Z.; Xing, W.; Lu, D. MIGT: Multi-modal image inpainting guided with text. *Neurocomputing* **2023**, *520*, 376–385. [CrossRef]
13. Pathak, D.; Krahenbuhl, P.; Donahue, J.; Darrell, T.; Efros, A.A. Context Encoders: Feature Learning by Inpainting. In Proceedings of the IEEE Conference on Computer Vision and Pattern Recognition (CVPR), Las Vegas, NV, USA, 27–30 June 2016.
14. Yang, C.; Lu, X.; Lin, Z.; Shechtman, E.; Wang, O.; Li, H. High-Resolution Image Inpainting Using Multi-Scale Neural Patch Synthesis. In Proceedings of the IEEE Conference on Computer Vision and Pattern Recognition (CVPR), Honolulu, HI, USA, 21–26 July 2017.
15. Yeh, R.A.; Chen, C.; Yian Lim, T.; Schwing, A.G.; Hasegawa-Johnson, M.; Do, M.N. Semantic Image Inpainting With Deep Generative Models. In Proceedings of the IEEE Conference on Computer Vision and Pattern Recognition (CVPR), Honolulu, HI, USA, 21–26 July 2017.
16. Iizuka, S.; Simo-Serra, E.; Ishikawa, H. Globally and Locally Consistent Image Completion. *ACM Trans. Graph.* **2017**, *36*, 1–14. [CrossRef]
17. Yu, J.; Lin, Z.; Yang, J.; Shen, X.; Lu, X.; Huang, T.S. Generative Image Inpainting with Contextual Attention. In Proceedings of the IEEE Conference on Computer Vision and Pattern Recognition (CVPR), Salt Lake City, UT, USA, 18–23 June 2018.
18. Liu, G.; Reda, F.A.; Shih, K.J.; Wang, T.C.; Tao, A.; Catanzaro, B. Image Inpainting for Irregular Holes Using Partial Convolutions. In Proceedings of the European Conference on Computer Vision (ECCV), Munich, Germany, 8–14 September 2018.
19. Yu, J.; Lin, Z.; Yang, J.; Shen, X.; Lu, X.; Huang, T.S. Free-Form Image Inpainting With Gated Convolution. In Proceedings of the IEEE/CVF International Conference on Computer Vision (ICCV), Seoul, Republic of Korea, 27 October–2 November 2019.
20. Yu, T.; Guo, Z.; Jin, X.; Wu, S.; Chen, Z.; Li, W.; Zhang, Z.; Liu, S. Region Normalization for Image Inpainting. In Proceedings of the AAAI Conference on Artificial Intelligence, New York, NY, USA, 7–12 February 2020; pp. 12733–12740.
21. Guo, X.; Yang, H.; Huang, D. Image Inpainting via Conditional Texture and Structure Dual Generation. In Proceedings of the IEEE/CVF International Conference on Computer Vision (ICCV), Virtual, 11–17 October 2021; pp. 14134–14143.
22. Liu, H.; Wan, Z.; Huang, W.; Song, Y.; Han, X.; Liao, J. Pd-gan: Probabilistic diverse gan for image inpainting. In Proceedings of the IEEE/CVF Conference on Computer Vision and Pattern Recognition, Nashville, TN, USA, 20–25 June 2021; pp. 9371–9381.
23. Wu, H.; Zhou, J.; Li, Y. Deep generative model for image inpainting with local binary pattern learning and spatial attention. *IEEE Trans. Multimed.* **2022**, *24*, 4016–4027. [CrossRef]
24. Mou, C.; Wang, Q.; Zhang, J. Deep generalized unfolding networks for image restoration. In Proceedings of the IEEE/CVF Conference on Computer Vision and Pattern Recognition, New Orleans, LA, USA, 18–24 June 2022; pp. 17399–17410.
25. Zhang, H.; Hu, Z.; Luo, C.; Zuo, W.; Wang, M. Semantic image inpainting with progressive generative networks. In Proceedings of the 26th ACM international conference on Multimedia, Seoul, Republic of Korea, 22–26 October 2018; pp. 1939–1947.
26. Li, J.; Wang, N.; Zhang, L.; Du, B.; Tao, D. Recurrent feature reasoning for image inpainting. In Proceedings of the IEEE/CVF Conference on Computer Vision and Pattern Recognition, Seattle, WA, USA, 13–19 June 2020; pp. 7760–7768.
27. Liu, H.; Jiang, B.; Xiao, Y.; Yang, C. Coherent semantic attention for image inpainting. In Proceedings of the IEEE/CVF International Conference on Computer Vision, Seoul, Republic of Korea, 27 October–2 November 2019; pp. 4170–4179.
28. Zeng, Y.; Fu, J.; Chao, H.; Guo, B. Learning pyramid-context encoder network for high-quality image inpainting. In Proceedings of the IEEE/CVF Conference on Computer Vision and Pattern Recognition, Long Beach, CA, USA, 15–20 June 2019; pp. 1486–1494.
29. Phutke, S.S.; Kulkarni, A.; Vipparthi, S.K.; Murala, S. Blind Image Inpainting via Omni-Dimensional Gated Attention and Wavelet Queries. In Proceedings of the IEEE/CVF Conference on Computer Vision and Pattern Recognition, Seattle, DC, USA, 17–21 June 2023; pp. 1251–1260.
30. Liu, H.; Jiang, B.; Song, Y.; Huang, W.; Yang, C. Rethinking image inpainting via a mutual encoder-decoder with feature equalizations. In Proceedings of the Computer Vision–ECCV 2020: 16th European Conference, Glasgow, UK, 23–28 August 2020; Proceedings, Part II 16; Springer: Berlin/Heidelberg, Germany, 2020; pp. 725–741.
31. Ma, X.; Deng, Y.; Zhang, L.; Li, Z. A Novel Generative Image Inpainting Model with Dense Gated Convolutional Network. *Int. J. Comput. Commun. Control.* **2023**, *18*, 5088. [CrossRef]
32. Xiong, W.; Yu, J.; Lin, Z.; Yang, J.; Lu, X.; Barnes, C.; Luo, J. Foreground-aware image inpainting. In Proceedings of the IEEE/CVF Conference on Computer Vision and Pattern Recognition, Long Beach, CA, USA, 15–20 June 2019; pp. 5840–5848.
33. Nazeri, K.; Ng, E.; Joseph, T.; Qureshi, F.Z.; Ebrahimi, M. Edgeconnect: Generative image inpainting with adversarial edge learning. *arXiv* **2019**, arXiv:1901.00212.
34. Guo, Z.; Chen, Z.; Yu, T.; Chen, J.; Liu, S. Progressive image inpainting with full-resolution residual network. In Proceedings of the 27th ACM International Conference on Multimedia, Nice, France, 21–25 October 2019; pp. 2496–2504.
35. Chen, Y.; Hu, H. An improved method for semantic image inpainting with GANs: Progressive inpainting. *Neural Process. Lett.* **2019**, *49*, 1355–1367. [CrossRef]
36. Li, J.; He, F.; Zhang, L.; Du, B.; Tao, D. Progressive reconstruction of visual structure for image inpainting. In Proceedings of the IEEE/CVF International Conference on Computer Vision, Seoul, Republic of Korea, 27 October–2 November 2019; pp. 5962–5971.
37. Liao, L.; Xiao, J.; Wang, Z.; Lin, C.W.; Satoh, S. Guidance and evaluation: Semantic-aware image inpainting for mixed scenes. In Proceedings of the Computer Vision–ECCV 2020: 16th European Conference, Glasgow, UK, 23–28 August 2020; Proceedings, Part XXVII 16; Springer: Berlin/Heidelberg, Germany, 2020; pp. 683–700.
38. Shi, K.; Alrabeiah, M.; Chen, J. Progressive with Purpose: Guiding Progressive Inpainting DNNs Through Context and Structure. *IEEE Access* **2023**, *11*, 2023–2034. [CrossRef]

39. Liu, W.; Liu, B.; Du, S.; Shi, Y.; Li, J.; Wang, J. Multi-stage Progressive Reasoning for Dunhuang Murals Inpainting. *arXiv* **2023**, arXiv:2305.05902.
40. Li, X.; Guo, Q.; Lin, D.; Li, P.; Feng, W.; Wang, S. MISF: Multi-level interactive Siamese filtering for high-fidelity image inpainting. In Proceedings of the IEEE/CVF Conference on Computer Vision and Pattern Recognition, New Orleans, LA, USA, 18–24 June 2022; pp. 1869–1878.
41. Chen, Y.; Xia, R.; Zou, K.; Yang, K. RNON: Image inpainting via repair network and optimization network. *Int. J. Mach. Learn. Cybern.* **2023**, *14*, 2945–2961. [CrossRef]
42. Karras, T.; Aila, T.; Laine, S.; Lehtinen, J. Progressive Growing of GANs for Improved Quality, Stability, and Variation. In Proceedings of the International Conference on Learning Representations, Vancouver, BC, Canada, 30 April–3 May 2018.
43. Zhou, B.; Lapedriza, A.; Khosla, A.; Oliva, A.; Torralba, A. Places: A 10 million Image Database for Scene Recognition. *IEEE Trans. Pattern Anal. Mach. Intell.* **2017**, *40*, 1452–1464. [CrossRef]
44. Karimov, A.; Kopets, E.; Kolev, G.; Leonov, S.; Scalera, L.; Butusov, D. Image preprocessing for artistic robotic painting. *Inventions* **2021**, *6*, 19. [CrossRef]
45. Karimov, A.; Kopets, E.; Leonov, S.; Scalera, L.; Butusov, D. A Robot for Artistic Painting in Authentic Colors. *J. Intell. Robot. Syst.* **2023**, *107*, 34. [CrossRef]

Disclaimer/Publisher's Note: The statements, opinions and data contained in all publications are solely those of the individual author(s) and contributor(s) and not of MDPI and/or the editor(s). MDPI and/or the editor(s) disclaim responsibility for any injury to people or property resulting from any ideas, methods, instructions or products referred to in the content.

Article

Edge-Oriented Compressed Video Super-Resolution

Zheng Wang, Guancheng Quan and Gang He *

School of Telecommunications Engineering, Xidian University, Xi'an 710071, China; jackwu0630@gmail.com (Z.W.); gcquan@stu.xidian.edu.cn (G.Q.)
* Correspondence: ghe@xidian.edu.cn

Abstract: Due to the proliferation of video data in Internet of Things (IoT) systems, in order to reduce the data burden, most social media platforms typically employ downsampling to reduce the resolution of high-resolution (HR) videos before video coding. Consequently, the loss of detail and the introduction of additional artifacts seriously compromise the quality of experience (QoE). Recently, the task of compressive video super-resolution (CVSR) has garnered significant attention, aiming to simultaneously eliminate compression artifacts and enhance the resolution of compressed videos. In this paper, we propose an edge-oriented compressed video super-resolution network (EOCVSR), which focuses on reconstructing higher-quality details, to effectively address the CVSR task. Firstly, we devised a motion-guided alignment module (MGAM) to achieve precise bi-direction motion compensation in a multi-scale manner. Secondly, we introduced an edge-oriented recurrent block (EORB) to reconstruct edge information by combining the merits of explicit and implicit edge extraction. In addition, benefiting from the recurrent structure, the receptive field of EOCVSR can be enhanced and the features can be effectively refined without introducing additional parameters. Extensive experiments conducted on benchmark datasets demonstrate that our method surpasses the performance of state-of-the-art (SOTA) approaches in both quantitative and qualitative evaluations. Our approach can provide users with high-quality and cost-effective HR videos by integrating with sensors and codecs.

Keywords: compressed video super-resolution; edge-oriented; recurrent structure

1. Introduction

Nowadays, the explosion of high-resolution (HR) videos has surged with the rapid advancement of portable device sensors and 5G communication technology. Some popular social media platforms, such as TikTok and WeChat, often downsample HR videos before compression to minimize the expenses associated with data storage and transmission, especially when bandwidth and storage space are severely constrained. The decoded low-resolution videos need to be upsampled back to their original resolution to meet user requirements. However, ensuring a satisfactory quality of experience (QoE) is extremely challenging due to the information loss caused by downsampling and the introduction of compression artifacts.

Despite the significant achievements of deep learning techniques in video quality enhancement (VQE) [1–6] and video super-resolution (VSR) tasks [7–11], simply cascading two networks to upsample the LR compressed videos in two stages often fail to yield satisfactory results. The main reason is that the information of the two independent processes cannot be effectively collaborated. Recently, there has been a growing interest in the task of compressed video super-resolution (CVSR) [12–14], which sought to address both compression artifacts removal and resolution enhancement in compressed videos simultaneously. An end-to-end restoration-reconstruction deep neural network (RR-DnCNN) [12] was first proposed to entirely solve degradation from compression and downsampling. The authors utilized upsampling skip connections to pass the useful features extracted by restoration to reconstruction. He et al. introduced a novel model, named Feature

Multiplexing Video Super-Resolution for Compressed Video (FM-VSR) [13], to recover high-quality, high-resolution videos from low-resolution videos compressed with high compression rates. However, these methods lack targeted processing for high-frequency components, particularly edge components. Furthermore, they solely rely on the mean squared error (MSE) loss function during training, resulting in the inadequate reconstruction of details and overly smooth and blurry outputs. There is a strong demand for robust CVSR algorithms that collaborate with sensors and video codecs to produce exceptional HR videos.

To this end, we propose an edge-oriented compressed video super-resolution network (EOCVSR), which pays more attention to reconstructing high-quality details to effectively address the CVSR task. EOCVSR takes the target frame and its two temporally adjacent frames as inputs to fully leverage the information from multiple frames. Firstly, we devised a motion-guided alignment module (MGAM) to achieve precise bi-direction motion compensation. We analyzed the explicit optical flow between two frames in a multi-scale manner to generate an attention map. The offset of the deformable convolution [15] generated under the guidance of the attention map enables accurate implicit temporal alignment. Secondly, we proposed an edge-oriented recurrent block (EORB) to reconstruct edge information. We adopted several conventional filters and some learnable convolution kernels with specific shapes to derive edge information. By combining the merits of explicit and implicit edge extraction, we can perform a more targeted and high-quality reconstruction of high-frequency components. Furthermore, we designed a recurrent structure to enhance the receptive field and the performance of EOCVSR without introducing additional parameters. The edge-oriented loss function is also employed during training to boost performance. Extensive experiments conducted on benchmark datasets demonstrate that our method surpasses the performance of SOTA approaches in both quantitative and qualitative evaluations. The main contributions can be summarized as follows:

- We propose an edge-oriented compressed video super-resolution network (EOCVSR) to address the CVSR problem. By incorporating a structure that specifically processes edge information and introduces edge-related loss functions, EOCVSR is able to reconstruct richer details and output higher-quality frames.
- We propose a motion-guided alignment module (MGAM) to achieve precise bi-direction motion compensation. The utilization efficiency of temporal information is enhanced by employing explicit motion information to guide the generation of offsets for implicit temporal alignment.
- We propose an edge-oriented recurrent block (EORB) to reconstruct edge information. Combining the merits of explicit and implicit edge extraction enables the high-quality reconstruction of high-frequency components. In addition, a recurrent structure is also adopted to realize effective feature refinement.

2. Related Works

2.1. Video Quality Enhancement (VQE)

With the widespread dissemination of video content and the increasing demand for storage, improving the quality of compressed videos has become a crucial task. Traditional video coding techniques introduce various distortions and compression artifacts during video compression, leading to a degradation in video quality. Consequently, researchers have begun exploring the use of deep learning methods to enhance the visual quality of compressed videos. An early work in the field of deep learning-based elimination of coding artifacts is the Variable-filter-size Residue-learning CNN (VRCNN) [1]. VRCNN integrates convolutional neural networks (CNNs) into the in-loop filter of High-Efficiency Video Coding (HEVC) [16] intra-coding, leading to improved coding performance. Wang et al. [2] introduced a Deep CNN-based Auto-Decoder (DCAD) approach, which directly enhances the decoded frames without modifying the specific components of the decoder. Recognizing the distinction between intra-coding and inter-coding in HEVC, Yang et al. [3] proposed the Quality-Enhancement Convolutional Neural Network (QECNN). QECNN comprises two subnetworks, namely QECNN-I and QECNN-B, which are designed to enhance the compressed I frames and P/B frames, respectively. Due to the lack of utilization of adjacent frames, the enhancement performance of these single-frame methods is limited. Therefore, some multi-frame methods [4–6] have been developed to leverage the temporal relationships between frames. Yang et al. observed that frames captured within a short period exhibit high similarity, leading them to propose a multi-frame quality enhancement (MFQE) approach [4]. This approach comprises a peak quality frames (PQFs) detector and a multi-frame convolutional neural network (MF-CNN). Building upon this, MFQE 2.0 [5] introduces an improved PQF-detector based on LSTM and a lightweight architecture for the MF-CNN, aiming to achieve enhanced performance more efficiently. Additionally, a spatio-temporal deformable fusion scheme [6] is employed, utilizing deformable convolutions to leverage temporal information. However, these existing methods need to be followed by an upsampling process to be applied to the CVSR task. The independence of the two processes from each other leads to an incoherent enhancement of the compressed video.

2.2. Video Super-Resolution (VSR)

Video super-resolution aims to improve video quality and detail reconstruction by learning the spatial and temporal characteristics of the video and upgrading the low-resolution video to high resolution. VSRnet [7] is an extension of the image super-resolution algorithm SRCNN [17] to the video domain. The major improvement lies in the inclusion of motion estimation and motion compensation modules, where the input is transformed from a single frame to multiple frames. Jo et al. introduced a DUF network [11] that generates dynamic upsampling filters and a residual image, which are computed depending on the local spatio-temporal neighborhood of each pixel to avoid explicit motion compensation. Tian et al. proposed a temporally deformable alignment network (TDAN) [10] to adaptively align the reference frame and each supporting frame with a feature level without computing optical flow. Concurrently, EDVR [8] devised an innovative spatio-temporal attention fusion module, along with enhanced deformable convolutions, to effectively handle motion compensation. However, when applying these methods to the CVSR task, compression artifacts may become more pronounced, significantly impacting video quality.

2.3. Compressed Video Super-Resolution (CVSR)

Recently, there has been a growing interest in the task of compressed video super-resolution (CVSR), which seeks to address both compression artifact removal and resolution enhancement in compressed videos simultaneously. An end-to-end restoration-reconstruction deep neural network (RR-DnCNN) [12] was proposed, which uses the degradation-aware technique to effectively deal with video compression distortion and up-sampling degradation. In its extended version, RR-DnCNN v2 [14], the authors redesigned the network architecture by adopting a U-shaped form and incorporating upsampling skip connections. This architectural modification enables the propagation of valuable features captured during the restoration process to the subsequent reconstruction stage. He et al. proposed a one-stage video super-resolution network (FM-VSR) [13] to recover high-quality, high-resolution videos from low-resolution videos. They make full use of the internal correlation between the VQE and VSR tasks by adopting a feature multi-plexing mechanism. However, these methods exhibit limitations in their treatment of high-frequency components, specifically edge components, as they do not incorporate targeted processing techniques. Additionally, their reliance solely on the mean squared error (MSE) loss function during the training phase leads to suboptimal reconstruction of fine details and outputs that are excessively smooth and blurry.

3. The Proposed EOCVSR Approach
3.1. Overall Framework

The overall framework of our proposed EOCVSR approach is depicted in Figure 1. EOCVSR takes the target frame I_t and its two temporally adjacent frames I_{t-1} and I_{t+1} as the input. First of all, a parameter-shared feature extraction module transforms three input frames from pixel space to feature space. As shown in Figure 1, the following procedure can be divided into two stages: restoration and reconstruction. The restoration stage aims to eliminate the compression artifacts. To make full use of temporal multi-frame information, a motion-guided alignment module is employed to achieve precise bi-direction motion compensation. Then, we stack three edge-oriented recurrent blocks (EORBs) to refine the aligned feature. The intermediate outputs of EORBs in the restoration stage are fused by a convolution layer. The fused feature is fed into a feature reconstruction module to obtain a residual image, which will be added to I_t to generate the result of the restoration stage O_t^{res}.

The reconstruction stage, which takes the fused feature output by the restoration stage as the input, aims to enhance the resolution and reconstruct high-quality details. The pixel-shuffle operation is first adopted to upsample the input feature. Another three cascading EORBs extract edge information at a larger scale. In the reconstruction stage, the output of each EORB is integrated with the output of the EORB, which is in the same order in the restoration stage. Such internal correlation between the two stages improves the learning ability of EOCVSR. Same as in the restoration stage, we use a convolution layer to fuse the outputs of EORBs from the reconstruction stage. Finally, the result of the reconstruction stage O_t^{rec} can be obtained by adding the residual image reconstructed from the high-resolution fused feature to the result of upsampling I_t using the Bi-cubic operation. The function of our proposed EOCVSR can be formulated as

$$O_t^{res}, O_t^{rec} = \Phi(I_{t-1}, I_t, I_{t+1}|\theta), \quad (1)$$

where Φ is the proposed approach and θ is the parameters of the network. The symbols that will appear with the corresponding explanations are shown in Table 1.

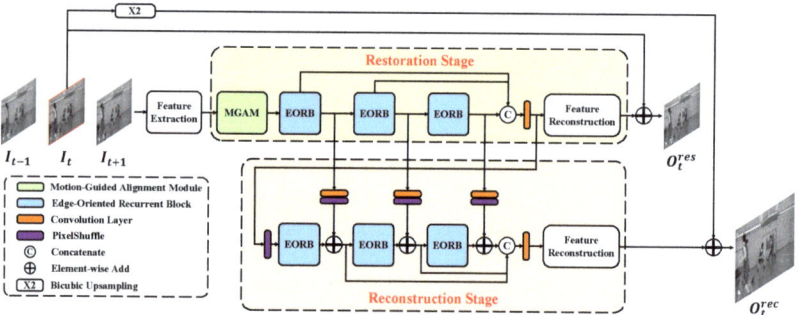

Figure 1. The overall framework of our proposed EOCVSR.

Table 1. List of symbols.

Symbols	Explanation
I_t	Input video frame t
O_t^{res}	Video restoration output
O_t^{rec}	Video reconstruction output
F_t	The feature of frame t
MV	Motion vector
O	The offset of the deformable convolution kernel
$F^{aligned}$	The aligned feature
F_t^{mgam}	The output feature of MGAM for frame t
$Conv_{m \times n}$	Convolutional layer with kernel of $m \times n$

Table 1. Cont.

Symbols	Explanation
$Softmax$	Softmax normalized activation function
$Sobel_{Dx}$	Vertical edge detection operator of Sobel
$Sobel_{Dy}$	Horizontal edge detection operator of Sobel
$Laplacian$	Laplacian edge detection operator
$\odot$	Element-wise product
$\oplus$	Element-wise add
G_t^{LR}	Low-resolution Ground Truth at frame t
G_t^{HR}	High-resolution Ground Truth at frame t
L_{mse}	Mean squared error loss function

3.2. Feature Extraction Module

The feature extraction module is defined as

$$F_n = f_{fe}(I_n), n \in \{t-1, t, t+1\}, \quad (2)$$

where $f_{fe}(\cdot)$ denotes the feature extraction function, and F represents the extracted feature of the input frames I. First, a convolution layer (Conv) is used to transform the input frames from a pixel space to a higher dimensional feature space. Then, we adopt a typical residual block (RB) [18] to roughly process the features. The parameters used to process the inputted three frames are shared.

3.3. Motion-Guided Alignment Module

The motion-guided alignment module (MGAM) aims to achieve precise bi-direction motion compensation and derive useful temporal multi-frame information for restoration. As shown in Figure 2, the structure of MGAM is divided into two parts, bi-direction motion-guided alignment and feature modulation. In the first part, we utilize explicit motion prior, i.e., optical flow, to guide feature-level temporal alignment. Take the forward alignment as an example, we first obtain the motion vector $MV_{t-1 \to t}$ using a pre-trained SpyNet [19], which is a widely used optical flow prediction network.

$$MV_{t-1 \to t} = SpyNet(I_t, I_{t-1}) \quad (3)$$

Then, we employ a parameter-shared Conv with a different dilation $d \in \{1, 2, 4\}$ to implement multi-scale analysis on $MV_{t-1 \to t}$. The analyzed results are concatenated, fused, and outputted by a softmax layer, resulting in the generation of a motion prior in the form of an attention map.

$$\begin{cases} h1 = Conv(MV_{t-1 \to t})_{d=1}, \\ h2 = Conv(MV_{t-1 \to t})_{d=2}, \\ h3 = Conv(MV_{t-1 \to t})_{d=3}, \\ MotionPrior = Softmax(Conv([h1, h2, h3])). \end{cases} \quad (4)$$

where $[\cdot, \cdot]$ and $Softmax$ are concatenation and the softmax layer. Meanwhile, an offset of the deformable convolution (DCN) kernel $O_{t-1 \to t}$ is dynamically predicted from F_t and F_{t-1}. Before directly applying $O_{t-1 \to t}$ to DCN, we fine-tune $O_{t-1 \to t}$ by multiplying the motion prior with it, so that it can better capture the motion correlation. By feeding F_{t-1} and the fine-tuned $O_{t-1 \to t}$ into a DCN, a forward-aligned feature $F_{t-1 \to t}^{aligned}$ can be obtained. The detailed process is given below:

$$\begin{cases} O_{t \to t-1} = Conv([F_t, F_{t-1}]), \\ F_{t-1 \to t}^{aligned} = DCN(F_{t-1}, O_{t \to t-1} \odot MotionPrior). \end{cases} \quad (5)$$

where DCN and $\odot$ denote the deformable convolution layer and element-wise product. The backward alignment takes F_t, F_{t+1}, and the backward optical flow takes $MV_{t+1\to t}$ as the input and outputs a backward-aligned feature $F_{t+1\to t}^{aligned}$. The processing is mirrored in the forward alignment.

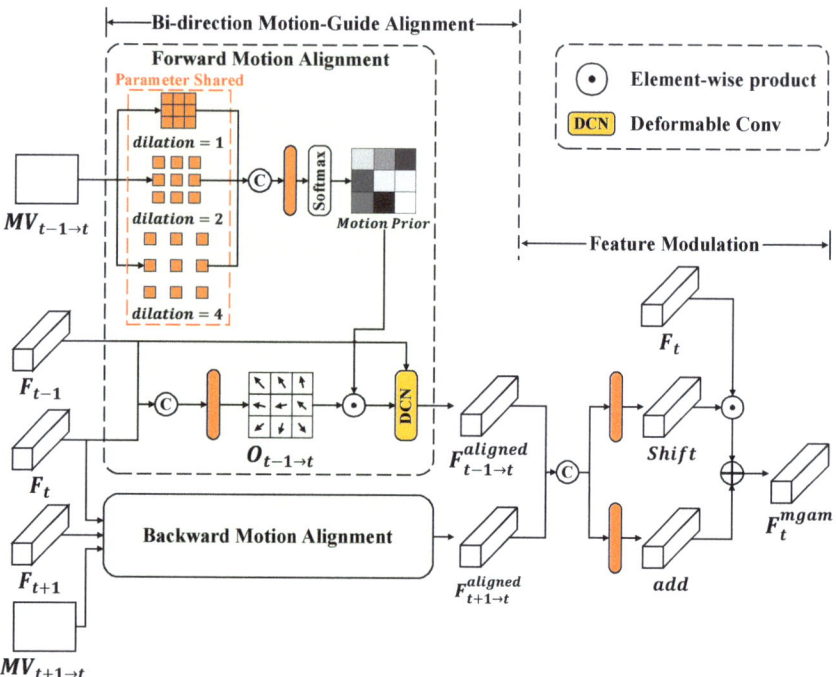

Figure 2. The structure of motion-guided alignment module (MGAM).

In the second part, we adopt the method of feature modulation to realize the fusion of multi-frame information. $F_{t-1\to t}^{aligned}$ and $F_{t+1\to t}^{aligned}$ are first concatenated together, and then the concatenation result is fed into two Convs to generate the shift and add weights for modulation, respectively. This operation can be described as

$$\begin{cases} shift = Conv([F_{t-1\to t}^{aligned}, F_{t+1\to t}^{aligned}]), \\ add = Conv([F_{t-1\to t}^{aligned}, F_{t+1\to t}^{aligned}]), \\ F_t^{mgam} = F_t \odot shift + add. \end{cases} \quad (6)$$

Thanks to precise motion alignment in the MGAM, the aligned features provide significant improvement to the overall performance.

3.4. Edge-Oriented Recurrent Block

Most existing CVSR approaches lack targeted processing for high-frequency components, particularly edge components. Furthermore, they solely rely on mean squared error (MSE) loss function during training, resulting in inadequate reconstruction of details and overly smooth and blurry outputs. To this end, we devise an edge-orient recurrent block (EORB) to augment the network's capacity for perceiving and reconstructing details. As shown in Figure 3, the recurrent unit is the key component of the EORB. The input of the recurrent unit is first processed by six different filters for edge-aware. To perceive horizontal edges, we adopt a horizontal Sobel filter and a learnable Conv with the size of 1×9. To perceive vertical edges, we adopt a vertical Sobel filter and another learnable

9×1 Conv. We also extract the second-order spatial derivative using a Laplacian filter, collaborating with a learnable 3×3 Conv. Combining the merits of pre-defined and learnable edge filters, the edge information can be efficiently derived. It is worth noting that the shapes of the six detectors will not change, and the parameters of the three learnable edge detectors can be updated through the back-propagation operation. Then, the summation of all the outputs, followed by the application of a global average pooling (GAP) layer, two Convs, and a softmax layer, generates the weights corresponding to each output. By multiplying the outputs with their corresponding weights and subsequently accumulating them, valuable edge information is filtered and preserved. The process within the recurrent unit is described as

$$\begin{cases} O_1 = Conv_{9\times1}(F_t^{RU-k}), O_2 = Sobel_{Dx}(F_t^{RU-k}) \\ O_3 = Conv_{1\times9}(F_t^{RU-k}), O_4 = Sobel_{Dy}(F_t^{RU-k}) \\ O_5 = Conv_{3\times3}(F_t^{RU-k}), O_6 = Laplacian(F_t^{RU-k}) \\ W_1 \sim W_6 = Softmax(Conv(Conv(GAP(\sum_{i=1}^{6} O_i)))), \\ RU(\cdot) = Conv(\sum_{i=1}^{6} O_i \odot W_i) \end{cases} \quad (7)$$

where $RU(\cdot)$ represents the process of the recurrent unit. $k \in [1, K-1]$ stands for the current iteration of recursion and K is the maximum number of recursions. At last, we stack three RBs to further process the feature. The output of the previous RB is utilized as the input of the next RB. Only the output of the last RB is added to the input feature of the EORB, which is F_t^{mgam}.

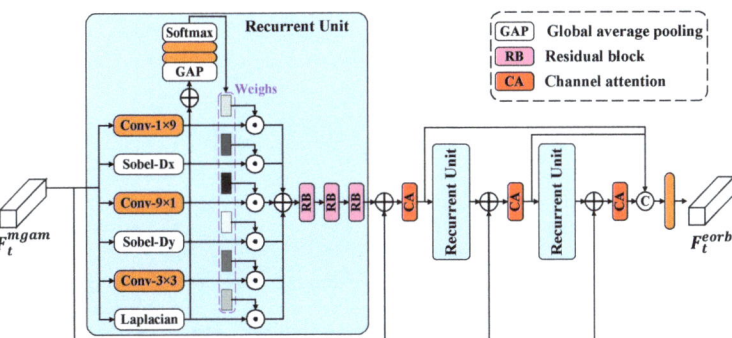

Figure 3. The structure of edge-oriented recurrent block (EORB).

To enhance the receptive field and performance without introducing additional parameters, we adopt a recurrent strategy. For the current recursion, the output of the previous recursion is added to the input of the EORB F_t^{input} and is subsequently passed through a channel attention layer. No matter how many times the recursion is performed during the training, it still belongs to the forward inference phase and does not involve updating the parameters. Therefore, in each recursion, the parameters of the recurrent unit are shared. The output of the EORB F_t^{eorb} is obtained by using a Conv to fuse the concatenation result of the outputs of all recursions. The detailed process is given below:

$$\begin{cases} F_t^{RU-k} = CA(RU(F_t^{input}) + F_t^{input}), \\ F_t^{RU-(k+1)} = CA(RU(F_t^{RU-k}) + F_t^{input}), \\ F_t^{eorb} = Conv([F_t^{RU-1}, F_t^{RU-(k+1)}]). \end{cases} \quad (8)$$

3.5. Feature Reconstruction Module

The feature reconstruction module takes the fused result of the outputs of all previous EORBs as the input. The feature reconstruction module consists of an RB and a Conv, which outputs a residual image R_t. The process is defined as

$$R_t = f_{fr}(Conv([F_t^{eorb-M}])), \qquad (9)$$

where $f_{fr}(\cdot)$ is the feature reconstruction function. M is the number of EORBs of each stage. In the restoration stage, the residual image will be added to I_t to obtain O_t^{res}. As for the reconstruction stage, the residual image will be added to the upsampling result of I_t to obtain O_t^{rec}. It is worth mentioning that the feature reconstruction modules in the restoration and reconstruction stages use different parameters.

3.6. Loss Function

Firstly, in addition to calculating the reconstruction error between the super-resolution result O_t^{rec} and the uncompressed HR frame G_t^{HR}, we also use the uncompressed LR frame G_t^{LR} as an auxiliary supervised label to assist the network training. Thus, the total loss function is defined as

$$L_{total} = \alpha * L_{res} + L_{rec}, \qquad (10)$$

where α is the weighted factor for the restoration loss.

However, only relying on the mean squared error (MSE) loss function during training may cause inadequate reconstruction of details and overly smooth and blurry outputs. To this end, we introduce an edge-orient loss function to direct constraints on the learning of edge reconstruction. Specifically, we explicitly extract high-frequency components using a Gaussian kernel blur $g(\cdot, \cdot)$ with the width of σ. The details of the loss function are shown below:

$$\begin{aligned} L_{res} &= L_{mse}(O_t^{res}, G_t^{LR}) + \beta * L_{mse}(OE_t^{res}, GE_t^{LR}) \\ L_{rec} &= L_{mse}(O_t^{rec}, G_t^{HR}) + \beta * L_{mse}(OE_t^{rec}, GE_t^{HR}) \\ OE_t &= O_t - g(O_t, \sigma = 3) \\ GE_t &= G_t - g(G_t, \sigma = 3) \end{aligned} \qquad (11)$$

4. Results

4.1. Experimental Setup

To train our proposed EOCVSR, we use the Vimeo dataset [19], which contains about 65,000 video sequences. Each sequence has seven frames with 448 × 256 resolution. We first use Bi-cubic interpolation to obtain uncompressed LR videos by a downsampling factor of two. Then, we compress these LR videos using FFmpeg [20] with the default mode and CRFs = 32, 37, 42, and 47. Finally, the bitstream is decoded to generate the compressed LR videos. To evaluate the trained EOCVSR, we adopt the test sequences from the standard HEVC common test condition (CTC) [21], the UVG dataset [22], and the MCL-JCV dataset [23]. The downsampling and compression settings are consistent with those in the training.

During the training, we randomly crop 120 × 120 patches from a mini-batch as the input. The batch size is set as eight. The hyperparameters regarding the network structure K and M are both set as three. The model trained with the loss function described in Section 3.5, and the weights α and β, are set as 0.2 and 0.1, respectively. The learning rate is initialized as 1×10^{-4} and then divided by a factor of 10 every 30 epochs. The training stops after 100 epochs. The Adam optimizer [23] is used by setting $\beta_1 = 0.9$ and $\beta_2 = 0.999$. During the evaluation, we use BD-BR [24], which presents the quality improvement (dB) at the same bitrate, and PSNR for quantitative analysis of the compressed video super-resolution results. All the models are implemented with PyTorch 1.4.0 [25] on Nvidia Geforce 2080Ti GPUs. All calculations are on the luminance channel (Y channel).

4.2. Performance of Proposed EOCVSR

The performance of our proposed EOCVSR is compared with the latest CVSR approaches, including RR-DnCNN v2 [14] and FM-VSR [13]. As shown in Table 2, we compare the characteristics between EOCVSR and other CVSR approaches. We also retrain some advanced video super-resolution approaches, including [8] and BasicVSR++ [9], over our training dataset for comparison. The comparison of the size of each model and computation cost is displayed in Table 3. We evaluate the quality enhancement, the rate–distortion performance, and subjective performance. The details are described as follows.

Table 2. Comparison with previous CVSR works.

Approach	Multi-Frame Utilization	Edge Preservation
RR-DnCNN v2	×	×
FM-VSR	✓	×
EOCVSR (proposed)	✓	✓

Table 3. The comparison of the number of model parameters and GFLOPs. GFLOPs are calculated on an image with an input size of 64 × 64.

	RR-DnCNN v2	FM-VSR	EDVR	BasicVSR++	EOCVSR
Parameter Number	1.8M	7.1M	2.7M	7.1M	3.5M
GFLOPs	20.9	96.1	66.6	104.3	88.4

4.2.1. Quality Enhancement

Table 4 presents the PSNR results over HEVC standard test sequences. First, the results illustrate that the proposed EOCVSR outperforms all the prior approaches over four compression ratios. Specifically, the PSNR of EOCVSR is 29.037 dB, which is 0.038 dB higher than the state-of-the-art approach, i.e., BasicVSR++ (28.999 dB), and 0.107–0.325 dB higher than others [8,13,14]. Table 5 shows the PSNR results over the UVG and MCL-JCV datasets, and the results demonstrate that EOCVSR also achieves the most significant boost. In terms of the UVG dataset, the PSNR improvement of EOCVSR ranges from 0.022 to 0.140. In terms of the MCL-JCV dataset, the PSNR improvement of EOCVSR ranges from 0.031 to 0.200. Thanks to MGAM's precise motion alignment and EORM's powerful edge information extraction and reconstruction capabilities, our proposed EOCVSR achieves the highest quality enhancement over all evaluation datasets.

Table 4. The comparison of PSNR gain over HEVC standard test sequences. Red indicates the best performance, and blue indicates the second-best.

QP	Class	Sequences	RR-DnCNN v2	FM-VSR	EDVR	BasicVSR++	EOCVSR
32	A	PeopleOnStreet	29.608	29.813	28.859	30.029	30.129
		Traffic	32.817	32.866	32.976	33.069	33.102
	B	BasketballDrive	31.726	31.902	32.422	32.471	32.531
		Cactus	31.012	31.297	31.566	31.712	31.735
		Kimono	34.456	34.735	34.799	34.844	34.876
		ParkScene	31.229	31.384	31.608	31.655	31.671
	C	BasketballDrill	30.177	30.379	30.460	30.636	30.706
		RaceHorses	24.544	26.111	27.524	27.599	27.609
	E	FourPeople	33.757	33.865	33.929	34.103	34.260
		Johnny	35.705	35.926	36.101	36.137	36.162
		KristenAndSara	34.282	34.510	34.660	34.919	34.973
		Average	32.017	32.072	32.348	32.470	32.523
37		Average	29.962	30.093	30.137	30.210	30.262

Table 4. *Cont.*

QP	Class	Sequences	RR-DnCNN v2	FM-VSR	EDVR	BasicVSR++	EOCVSR
42		Average	27.559	27.610	27.775	27.851	27.838
47		Average	25.308	25.366	25.459	25.464	25.524
		Overall	28.712	28.785	28.930	28.999	29.037

Table 5. The comparison of average PSNR gain over the UVG and MCL-JCV datasets. Red indicates the best performance, and blue indicates the second-best.

Dataset	QP	RR-DnCNN v2	FM-VSR	EDVR	BasicVSR++	EOCVSR
UVG	32	35.352	35.358	35.450	35.526	35.558
	37	33.142	33.167	33.151	33.237	33.250
	42	30.683	30.701	30.755	30.822	30.822
	47	28.267	28.243	28.326	28.331	28.375
	Overall	31.861	31.867	31.920	31.979	32.001
MCL-JCV	32	34.860	34.935	35.053	35.140	35.172
	37	32.748	32.749	32.774	32.890	32.930
	42	30.426	30.480	30.500	30.575	30.568
	47	28.222	28.304	28.313	28.328	28.386
	Overall	31.564	31.617	31.660	31.733	31.764

4.2.2. Rate–Distortion Performance

Here, we evaluate the rate–distortion of EOCVSR over three datasets, and the results are shown in Table 6. Considering full-resolution compression as an anchor, the performance has been improved in terms of 14.364% BD-BR reduction on average over HEVC standard test sequences, which is 0.843% more than BasicVSR++ (13.521%), and 2.737–8.277% more than other approaches [8,13,14]. As for the UVG and the MCL-JCV datasets, the bitrate saving achieves 29.849% and 28.217%. To present more intuitively, rate–distortion curves of our and other approaches over all test sequences are shown in Figure 4. In this figure, we can observe that the curve of EOCVSR is above that of full-resolution compression. In brief, the quantitative results demonstrate that our proposed EOCVSR realizes the best compression performance.

Table 6. The comparison of BD-BR over HEVC standard test sequences. Red indicates the best performance, and blue indicates the second-best.

Dataset	Class	Sequences	RR-DnCNN v2	FM-VSR	EDVR	BasicVSR++	EOCVSR
HEVC	A	PeopleOnStreet	−11.621	−13.780	−15.541	−18.113	−20.621
		Traffic	−3.916	−5.019	−6.153	−7.833	−8.584
	B	BasketballDrive	0.687	−7.133	−11.365	−14.361	−14.140
		Cactus	−0.224	−9.965	−12.462	−14.842	−15.386
		Kimono	−13.743	−17.421	−20.907	−22.182	−22.501
		ParkScene	5.323	−4.408	−7.121	−8.035	−8.452
	C	BasketballDrill	0.307	−3.147	−4.883	−6.685	−8.300
		RaceHorses	−6.599	−7.366	−8.185	−10.610	−11.420
	E	FourPeople	−11.528	−12.154	−12.788	−13.591	−14.836
		Johnny	−19.018	−19.580	−20.076	−21.614	−22.046
		KristenAndSara	−7.181	−7.877	−8.414	−10.864	−11.721
		Average	−6.137	−9.804	−11.627	−13.521	−14.364
UVG		Average	−27.086	−27.313	−27.804	−29.502	−29.849
MCL-JCV		Average	−24.066	−24.526	−24.913	−27.599	−28.217

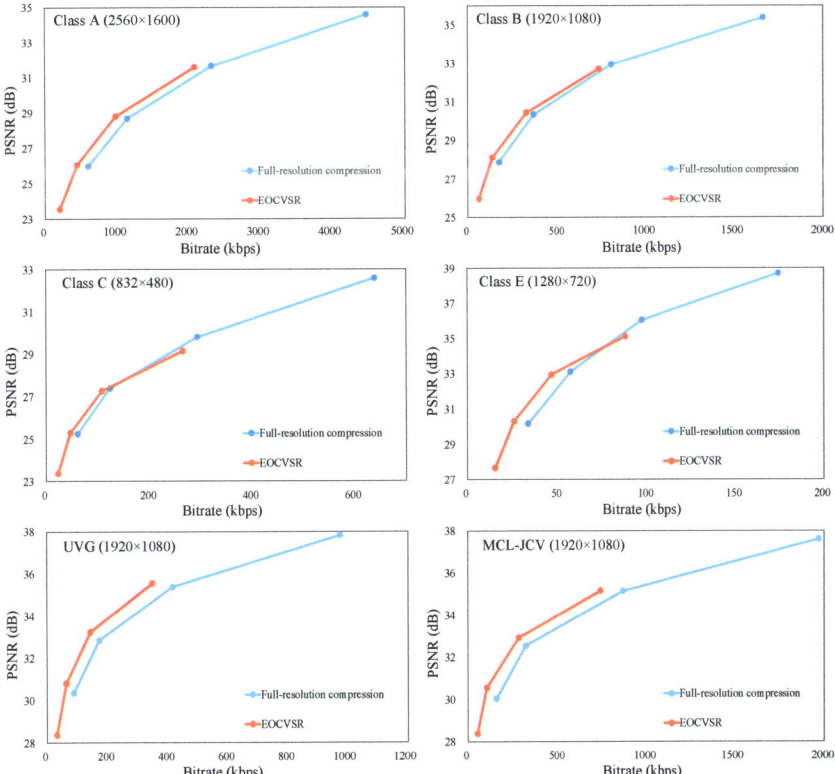

Figure 4. The RD curve over all test datasets of various resolutions under different compression distortion conditions.

4.2.3. Subjective Performance

Figure 5 shows the visual results of the different methods. Observing the results, it is evident that our proposed EOCVSR has demonstrated substantial advancements in visual perception when compared to other approaches. The compression artifacts are effectively eliminated. Thanks to the excellent edge-awareness of EOCVSR, the edges and textures are reconstructed with high quality. For example, the face in the sequence BasketballDrive and the letters in the sequence KristenAndSara. Therefore, our EOCVSR approach achieves promising performance in subjective quality.

4.3. Ablation Study

In this section, we conduct several ablation experiments to analyze the impact of different structures of the proposed EOCVSR on performance. All the experiments are performed over the MCL-JCV dataset and the compression ratio CRF is set as 47. In general, the results demonstrate that EOCVSR is delicately designed to achieve optimal performance. The details are described as follows.

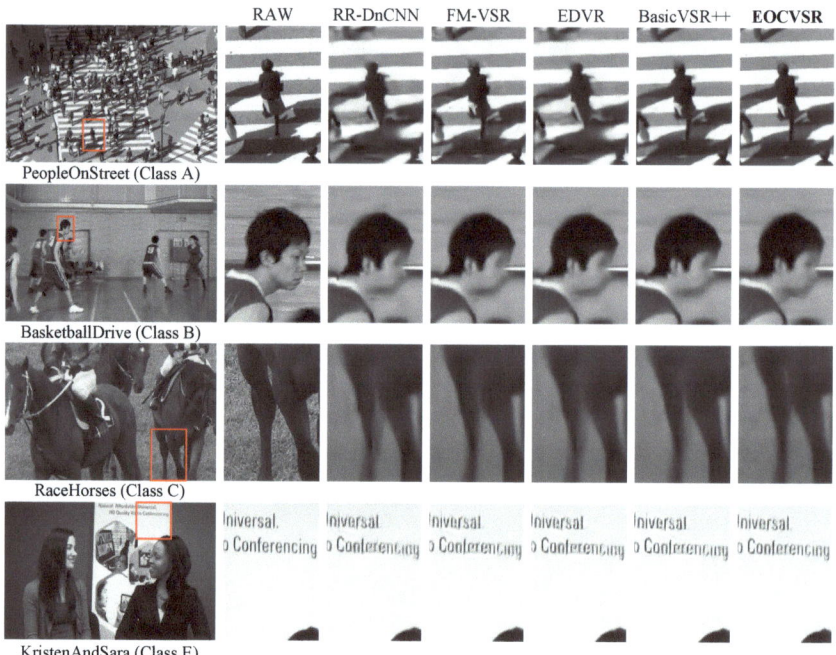

Figure 5. Visual comparisons of different methods on video frames from HEVC standard test sequences (QP = 42). The zoom-in of red box area is shown.

4.3.1. Analysis of the EORB

An ablation study has been conducted to verify the effectiveness of our proposed EORB. We modify the EORB by removing the six filters used to perceive the edges and leaving only the RBs. As shown in Table 7, the original EORB achieves a 28.386 dB improvement in PSNR, while the performance of the modified EORB degrades to 28.317 dB. The results prove that combining the merits of pre-defined and learnable edge filters enables EOCVSR to efficiently extract the edge information for higher performance. Furthermore, we provide a feature map visualization example. As shown in Figure 6, the original EORB is more sensitive to edges and textures.

Table 7. The performance of the EORB with edge-perceiving filters vs. without edge-perceiving filters.

	EORB w/o Edge-Perceiving Filters	EORB w/ Edge-Perceiving Filters
PSNR (dB)	28.317	28.386

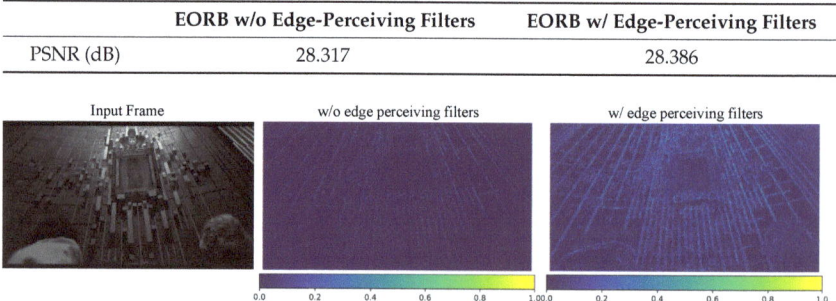

Figure 6. The feature map output by EORB with edge-perceiving filters vs. without edge-perceiving filters.

4.3.2. Analysis of the Number of Recursions K

We design a recurrent structure to enhance the receptive field and the performance of EOCVSR without introducing additional parameters, and the number of recursions within an EORB has an impact on the performance of EOCVSR. As shown in Figure 7a, the performance increases quickly when *K* is from one to three, while the performance increases slowly when *K* is larger than three. Considering more recursions may extend the processing time, *K* is set as three in this work.

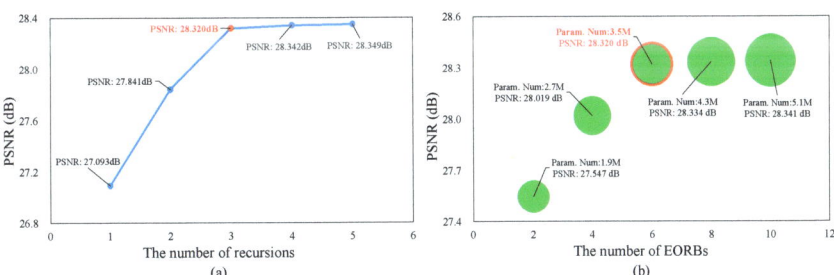

Figure 7. Ablation study on (**a**) the number of recursions within an EORB and (**b**) the number of EORBs.

4.3.3. Analysis of the Number of EORBs *M*

We evaluate EOCVSR with different numbers of EORBs. Integrating more EORBs can enhance the representation capabilities of the network but introduce more parameters. Figure 7b shows the performance of quality enhancement versus the number of EORBs *M*. The performance gain increases slowly when *M* is from 6 to 10. Thus, *M* is set as six (three in the restoration stage and another three in the reconstruction stage) in this work.

4.3.4. Model Adaption

We evaluate the adaptability of the proposed EOCVSR to different types of tasks. Theoretically, the proposed design can be applied to many other video tasks, such as denoising tasks, enhancement tasks, deblurring tasks, and so on. We add Gaussian noise to the image frames to train and test the ability of EOCVSR to perform super-resolution denoising at noise levels of σ = 15, 25, and 50. The parameter settings and network training are the same as for the original task. We compared the performance of EOCVSR with MIRNet [26]. As shown in Table 8, the PSNR of EOCVSR is improved by 0.45–1.4 dB. The results show that our method outperforms the learning-based super-resolution denoising method, illustrating that our model possesses the ability to adapt to different types of tasks.

Table 8. The adaption of the proposed design on video super-resolution denoising task.

Noise Level	Scale	MIRNet	EOCVSR
$\sigma = 15$	×2	34.65	35.10
$\sigma = 25$	×2	33.86	34.45
$\sigma = 50$	×2	31.06	32.46

5. Conclusions

In this paper, we proposed an edge-oriented compressed video super-resolution network (EOCVSR), which pays more attention to reconstructing high-quality details to effectively address the CVSR task. We devised a motion-guided alignment module (MGAM), which uses the explicit optical flow to direct the generation of offsets for deformable convolutions leading to precise bi-direction motion compensation. In addition, we proposed an edge-oriented recurrent block (EORB) to reconstruct edge information. Several pre-defined filters and some learnable convolution kernels with specific shapes were exploited to derive edge information. By combining the merits of explicit and implicit edge extraction, we can perform a more targeted and high-quality reconstruction of high-frequency components.

We also designed a recurrent structure to enhance the receptive field and the performance of EOCVSR without introducing additional parameters. Our approach can provide users with high-quality and cost-effective HR videos by integrating with sensors and codecs.

Author Contributions: Conceptualization, Z.W. and G.H.; methodology, Z.W.; software, Z.W.; validation, G.Q.; formal analysis, Z.W.; investigation, G.H.; resources, G.H.; data curation, G.Q.; writing—original draft preparation, Z.W. and G.Q.; writing—review and editing, G.H.; visualization, Z.W. and G.Q.; supervision, G.H.; project administration, G.H.; funding acquisition, G.H. All authors have read and agreed to the published version of this manuscript.

Funding: This work is supported by project CEIEC-2022-ZM02-0247.

Institutional Review Board Statement: Not applicable.

Informed Consent Statement: Not applicable.

Data Availability Statement: The data presented in this study are available on request from the corresponding author.

Conflicts of Interest: The authors declare no conflicts of interest.

References

1. Dai, Y.; Liu, D.; Wu, F. A convolutional neural network approach for post-processing in HEVC intra coding. In Proceedings of the MultiMedia Modeling: 23rd International Conference, MMM 2017, Reykjavik, Iceland, 4–6 January 2017; Proceedings, Part I 23; Springer: Berlin/Heidelberg, Germany, 2017; pp. 28–39.
2. Wang, T.; Chen, M.; Chao, H. A novel deep learning-based method of improving coding efficiency from the decoder-end for HEVC. In Proceedings of the 2017 Data Compression Conference (DCC), Snowbird, UT, USA, 4–7 April 2017; pp. 410–419.
3. Pan, Z.; Yi, X.; Zhang, Y.; Jeon, B.; Kwong, S. Efficient in-loop filtering based on enhanced deep convolutional neural networks for HEVC. *IEEE Trans. Image Process.* **2020**, *29*, 5352–5366. [CrossRef] [PubMed]
4. Yang, R.; Xu, M.; Wang, Z.; Li, T. Multi-frame quality enhancement for compressed video. In Proceedings of the IEEE Conference on Computer Vision and Pattern Recognition, Salt Lake City, UT, USA, 18–23 June 2018; pp. 6664–6673.
5. Guan, Z.; Xing, Q.; Xu, M.; Yang, R.; Liu, T.; Wang, Z. MFQE 2.0: A new approach for multi-frame quality enhancement on compressed video. *IEEE Trans. Pattern Anal. Mach. Intell.* **2019**, *43*, 949–963. [CrossRef] [PubMed]
6. Deng, J.; Wang, L.; Pu, S.; Zhuo, C. Spatio-temporal deformable convolution for compressed video quality enhancement. In Proceedings of the AAAI Conference on Artificial Intelligence, New York, NY, USA, 7–12 February 2020; Volume 34, pp. 10696–10703.
7. Kappeler, A.; Yoo, S.; Dai, Q.; Katsaggelos, A.K. Video super-resolution with convolutional neural networks. *IEEE Trans. Comput. Imaging* **2016**, *2*, 109–122. [CrossRef]
8. Wang, X.; Chan, K.C.; Yu, K.; Dong, C.; Change Loy, C. Edvr: Video restoration with enhanced deformable convolutional networks. In Proceedings of the IEEE/CVF Conference on Computer Vision and Pattern Recognition Workshops, Long Beach, CA, USA, 16–20 June 2019.
9. Chan, K.C.; Zhou, S.; Xu, X.; Loy, C.C. Basicvsr++: Improving video super-resolution with enhanced propagation and alignment. In Proceedings of the IEEE/CVF Conference on Computer Vision and Pattern Recognition, New Orleans, LA, USA, 18–24 June 2022; pp. 5972–5981.
10. Tian, Y.; Zhang, Y.; Fu, Y.; Xu, C. Tdan: Temporally-deformable alignment network for video super-resolution. In Proceedings of the IEEE/CVF Conference on Computer Vision and Pattern Recognition, Seattle, WA, USA, 14–19 June 2020; pp. 3360–3369.
11. Jo, Y.; Oh, S.W.; Kang, J.; Kim, S.J. Deep video super-resolution network using dynamic upsampling filters without explicit motion compensation. In Proceedings of the IEEE Conference on Computer Vision and Pattern Recognition, Salt Lake City, UT, USA, 18–23 June 2018; pp. 3224–3232.
12. Ho, M.M.; He, G.; Wang, Z.; Zhou, J. Down-sampling based video coding with degradation-aware restoration-reconstruction deep neural network. In Proceedings of the MultiMedia Modeling: 26th International Conference, MMM 2020, Daejeon, Republic of Korea, 5–8 January 2020; Proceedings, Part I 26; Springer: Berlin/Heidelberg, Germany, 2020; pp. 99–110.
13. He, G.; Wu, S.; Pei, S.; Xu, L.; Wu, C.; Xu, K.; Li, Y. FM-VSR: Feature Multiplexing Video Super-Resolution for Compressed Video. *IEEE Access* **2021**, *9*, 88060–88068. [CrossRef]
14. Ho, M.M.; Zhou, J.; He, G. RR-DnCNN v2. 0: Enhanced restoration-reconstruction deep neural network for down-sampling-based video coding. *IEEE Trans. Image Process.* **2021**, *30*, 1702–1715. [CrossRef]
15. Dai, J.; Qi, H.; Xiong, Y.; Li, Y.; Zhang, G.; Hu, H.; Wei, Y. Deformable convolutional networks. In Proceedings of the IEEE International Vonference on Vomputer Vision, Venice, Italy, 22–29 October 2017; pp. 764–773.
16. Sullivan, G.J.; Ohm, J.R.; Han, W.J.; Wiegand, T. Overview of the high efficiency video coding (HEVC) standard. *IEEE Trans. Circuits Syst. Video Technol.* **2012**, *22*, 1649–1668. [CrossRef]

17. Dong, C.; Loy, C.C.; Tang, X. Accelerating the super-resolution convolutional neural network. In Proceedings of the Computer Vision–ECCV 2016: 14th European Conference, Amsterdam, The Netherlands, 11–14 October 2016; Proceedings, Part II 14; Springer: Berlin/Heidelberg, Germany, 2016; pp. 391–407.
18. Lim, B.; Son, S.; Kim, H.; Nah, S.; Mu Lee, K. Enhanced deep residual networks for single image super-resolution. In Proceedings of the IEEE Conference on Computer Vision and Pattern Recognition Workshops, Honolulu, HI, USA, 21–26 July 2017; pp. 136–144.
19. Ranjan, A.; Black, M.J. Optical flow estimation using a spatial pyramid network. In Proceedings of the IEEE Conference on Computer Vision and Pattern Recognition, Honolulu, HI, USA, 21–26 July 2017; pp. 4161–4170.
20. Newmarch, J.; Newmarch, J. Ffmpeg/libav. *Linux Sound Program.* **2017**, *19*, 227–234.
21. Bossen, F. Common test conditions and software reference configurations. *JCTVC-L1100* **2013**, *12*, 1.
22. Mercat, A.; Viitanen, M.; Vanne, J. UVG dataset: 50/120fps 4K sequences for video codec analysis and development. In Proceedings of the 11th ACM Multimedia Systems Conference, Istanbul, Turkey, 8–11 June 2020; pp. 297–302.
23. Wang, H.; Gan, W.; Hu, S.; Lin, J.Y.; Jin, L.; Song, L.; Wang, P.; Katsavounidis, I.; Aaron, A.; Kuo, C.C.J. MCL-JCV: A JND-based H. 264/AVC video quality assessment dataset. In Proceedings of the 2016 IEEE international conference on image processing (ICIP), Phoenix, AZ, USA, 25–28 September 2016; pp. 1509–1513.
24. Grois, D.; Marpe, D.; Mulayoff, A.; Itzhaky, B.; Hadar, O. Performance comparison of h. 265/mpeg-hevc, vp9, and h. 264/mpeg-avc encoders. In Proceedings of the 2013 Picture Coding Symposium (PCS), San Jose, CA, USA, 8–11 December 2013; pp. 394–397.
25. Imambi, S.; Prakash, K.B.; Kanagachidambaresan, G. PyTorch. In *Programming with TensorFlow: Solution for Edge Computing Applications*; Springer: Berlin/Heidelberg, Germany, 2021; pp. 87–104.
26. Zamir, S.W.; Arora, A.; Khan, S.; Hayat, M.; Khan, F.S.; Yang, M.H.; Shao, L. Learning Enriched Features for Real Image Restoration and Enhancement. *arXiv* **2020**, arXiv:2003.06792v2.

Disclaimer/Publisher's Note: The statements, opinions and data contained in all publications are solely those of the individual author(s) and contributor(s) and not of MDPI and/or the editor(s). MDPI and/or the editor(s) disclaim responsibility for any injury to people or property resulting from any ideas, methods, instructions or products referred to in the content.

Article

Transformer-Based Multiple-Object Tracking via Anchor-Based-Query and Template Matching

Qinyu Wang, Chenxu Lu, Long Gao * and Gang He

State Key Laboratory of Integrated Service Networks, School of Telecommunications Engineering, Xidian University, No. 2, South Taibai Street, Hi-Tech Development Zone, Xi'an 710071, China; lcx031478@gmail.com (Q.W.); 22011210922@stu.xidian.edu.cn (C.L.); ghe@xidian.edu.cn (G.H.)
* Correspondence: lgao@xidian.edu.cn; Tel.: +86-13772054600

Abstract: Multiple object tracking (MOT) plays an important role in intelligent video-processing tasks, which aims to detect and track all moving objects in a scene. Joint-detection-and-tracking (JDT) methods are thriving in MOT tasks, because they accomplish the detection and data association in a single stage. However, the slow training convergence and insufficient data association limit the performance of JDT methods. In this paper, the anchor-based query (ABQ) is proposed to improve the design of the JDT methods for faster training convergence. By augmenting the coordinates of the anchor boxes into the learnable queries of the decoder, the ABQ introduces explicit prior spatial knowledge into the queries to focus the query-to-feature learning of the JDT methods on the local region, which leads to faster training speed and better performance. Moreover, a new template matching (TM) module is designed for the JDT methods, which enables the JDT methods to associate the detection results and trajectories with historical features. Finally, a new transformer-based MOT method, ABQ-Track, is proposed. Extensive experiments verify the effectiveness of the two modules, and the ABQ-Track surpasses the performance of the baseline JDT methods, TransTrack. Specifically, the ABQ-Track only needs to train for 50 epochs to achieve convergence, while that for TransTrack is 150 epochs.

Keywords: video processing; multiple-object tracking; anchor-based query; transformer; template matching

---

Citation: Wang, Q.; Lu, C.; Gao, L.; He, G. Transformer-Based Multiple-Object Tracking via Anchor-Based-Query and Template Matching. *Sensors* **2024**, *24*, 229. https://doi.org/10.3390/s24010229

Academic Editor: Shih-Chia Huang

Received: 20 November 2023
Revised: 14 December 2023
Accepted: 22 December 2023
Published: 30 December 2023

Copyright: © 2023 by the authors. Licensee MDPI, Basel, Switzerland. This article is an open access article distributed under the terms and conditions of the Creative Commons Attribution (CC BY) license (https://creativecommons.org/licenses/by/4.0/).

1. Introduction

Multiple object tracking (MOT) is a thriving field in intelligent video processing, and has wide applications in autonomous vehicles, video surveillance, and intelligent transportation [1]. The objective of the MOT methods is to estimate the movement states of objects and maintain identifications of those objects within a single video stream. Despite the numerous efforts dedicated to MOT, the design of efficient and robust trackers remains a challenge, primarily due to the need to address two critical sub-tasks within a single tracker: detection and data association [2].

In the field of Multiple Object Tracking (MOT), Tracking-by-Detection (TBD) has been the domain framework, which utilized detection methods to find objects in the scenes, and applied the data association to join the detection results into the trajectories. The detection and data association were conducted separately, which neglected the inner link between the two procedures [3]. Furthermore, the training of the TBD methods was complicated since the training procedures of models for detection and data association were different [4,5]. To overcome these problems, Joint-Detection-and-Tracking (JDT) methods were proposed, which performed the detection and data association in one stage [6–8]. Specifically, the JDT modified the transformer-based network for MOT, which consists of the encoder and the decoder. The encoder is utilized for feature enhancement, and the decoder is implemented for detecting the objects based on the object queries and the tracker queries, respectively. As shown in Figure 1, the decoder detects the objects in the images by applying the object queries, which are learned in the training procedure. Meanwhile, the queries obtained in the previous image (track queries) were fed into the decoder, which shared the same structure with the decoder for detection, to estimate the location of the object in different trajectories. By computing the IoU scores of the detection results based on the two set of queries, the identifications were maintained. In general, the JDT methods unified the detection and data association in one network, and competitive performance was achieved.

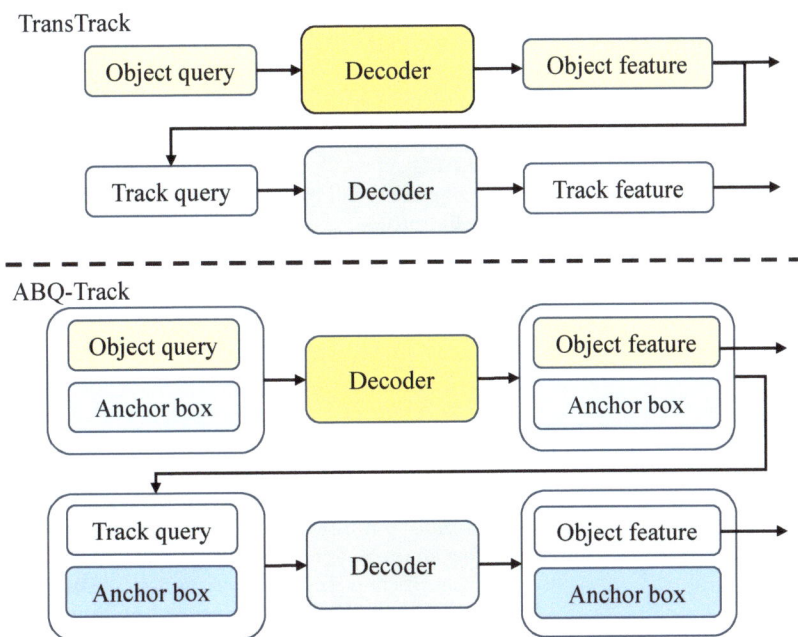

Figure 1. Comparison of the pipeline of the classical transformer-based MOT tracker, TransTrack, and the proposed tracker, ABQ-Track.

However, the JDT methods face challenges related to extended training times due to inappropriate design of the learnable queries [9]. In the cross-attention module of the decoder, an attention map is learned with the queries and keys. The two queries, the object queries and the track queries, are applied to match the content and spatial information in the keys [10], and the spatial information in the queries is not encoded with the same pattern [9]. Consequently, more training epochs are needed for the JDT methods to learn strong enough queries. Additionally, JDT associates detection results with tracking results from previous frames with spatial information, the IoU scores, which is insufficient for maintaining stable trajectories. Objects within a trajectory may become occluded by other objects in the background, and their appearance may undergo significant changes [11,12]. The data association methods in JDT lack an explicit model for maintaining stable trajectories.

To address the above issues, a new MOT method, ABQ-Track, is proposed. To begin with, the explicit prior spatial knowledge for each trajectory is introduced by 4D anchor box (ABQ) and added into each query corresponding to the trajectory. The 4D anchor boxes consist of the locations and sizes of objects in the previous image, helping the data association module focus on a local region corresponding to the object in each trajectory, and the locations and sizes are encoded with the same ways as the positional embedding for the keys. Consequently, the tracker augmented with additional spatial prior knowledge achieves better performance, while the network is trained for fewer epochs. Additionally, the template matching (TM) is implemented to stabilize the trajectories. In addition to associating the trajectories and the detection results with spatial information, the TM enables the JDT to accomplish the data association with appearance information. In previous images, the TM reserves the appearances of the objects in different trajectories and fuses these appearances to obtain templates, which represent the appearances of objects in different trajectories. By comparing the templates with the features corresponding to the detection results, the ABQ-Track associates the data with appearance information. Finally, the ABQ and TM are instanced in the classical JDT method, TransTrack [7], as shown in Figure 1. Extensive experiments verify the effectiveness of the proposed methods. ABQ-Track achieves better performance than TransTrack on MOT-17 [13] and MOT-20 [14], and only needs to be trained for 50 epochs, which is much faster than that of TransTrack.

In general, the main contributions of this work can be summarized as follows:

(1) The ABQ approach is proposed to reduce the convergence time and improve the discriminative ability of the JDT method by incorporating explicit spatial knowledge into the queries.
(2) The TM method is introduced to stabilize the trajectories by associating the trajectories and detection results with historical appearance information in the trajectories.
(3) A new JDT method based on the two methods, ABQ-Track, is proposed. Extensive experiments are conducted, and the results verify the effectiveness of the proposed methods. Moreover, the ABQ-Track surpasses the performance of the other JDT methods, TransTrack, in faster training convergence.

2. Related Work

With the development of intelligent sensors, intelligent video-processing methods have been thriving for decades. Multi-object tracking (MOT) is an important task in intelligent video processing owing to its wide applications in the real-world. MOT focuses on tracking an unknown number of objects within given categories [15]. In this section, the two main pipelines, tracking-by-detection (TBD) and joint-detection-and-tracking (JDT), are reviewed firstly. The transformer architecture is then introduced since the method proposed in this paper is based on it.

2.1. Tracking-by-Detection

TBD methods directly applied the given detection results in the MOT datasets or the detection results outputted by existing detectors [16–19], and focused on associating the detection results with the trajectories across frames in the image sequences [3,20,21]. The simple online and real-time tracking (SORT) associated the detection results by computing the distance between the locations of the currently detected boxes and the predicted boxes via Hungarian algorithm [20]. Predicted boxes were estimated with Kalman Filter (KF) [22]. Only associating the data with positional information was insufficient, which caused unstable trajectories [23]. DeepSORT further augmented a metric based on the similarity of features in the data association method [21]. Following the two works, efforts have been devoted into exploring more discriminative metrics on features and positional information [23–27]. For obtaining better predicted boxes, some methods merged the detection scores into the KF [23,24] or adopted the camera motion compensation (CMC) [25–27]. For better association based on the features, additional neural networks for feature extraction was applied to obtain the similarity metrics for the features [2,28,29]. In general, a large number of TBD methods have been proposed, and leading performance has been achieved. Nevertheless, the high performance of the TBD methods relied on strong detectors and complex models, which restricted their applications.

2.2. Joint-Detection-and-Tracking

The joint-detection-and-tracking (JDT) methods aimed to detect and track multiple objects in one stage. The pioneering works conducted the object detection in current frame, and performed the data association on two successive frames [30,31]. The CenterTrack then took the objects as points, and performed a tracking-conditional detector to detect objects and associate the data [32]. JDE [2] and FairMOT [33] built object detection networks and ReID network with shared feature extraction networks. Moreover, as Transformer [34] has been introduced in computer vision and impressive performance was achieved, it has been applied in MOT [7,8]. TransTrack built the MOT network with a detection network to obtain the detection boxes with object queries, and augmented an additional decoder for estimating the predicted boxes based on track queries [7]. By computing the IoU of two sets of boxes, it associated the detection boxes and predicted boxes, which was corresponding to the trajectories. TrackFormer directly applied the track queries from the previous frame as the inputs of the decoder in the current frame, and detected the objects with both locations and identities [8].

2.3. Transformer-Based MOT Method

Transformer has been a popular architecture in computer vision [34], and significant performance improvements have been achieved in numerous tasks, i.e., classification [35,36], detection [37,38], and segmentation [39]. The Vision Transformer (ViT) introduced a pure transformer network for computer vision tasks, which flatten the images into tokens and process them with stacked layers of multi-head self-attention [40]. The performance improvements were gained for the global context modeling ability of the transformer. However, the local information was neglected and the computational cost was high. Some works proposed designs into ViT to learn local context information [36,41]. Others devoted to reduce the computation and memory cost of the vanilla multi-head self-attention [42–44]. For the detection task, DETR [37] built the detection network with encoder–decoder architecture based on transformer, and utilized learnable queries to predict objects without non-maximum suppression. Many follow-up studies have explored the methods to address the slow convergence of DETR by reforming the learnable queries [10,45,46].

3. Proposed Methodology

In this section, the mechanism of a representative JDT method, TransTrack, is revisited firstly, where the proposed methods are implemented. The anchor-based queries are then presented for the JDT methods. Following this, the template matching is sketched. Finally, the training and interfering of the new tracker based on the proposed modules are detailed.

3.1. Revisit of the TransTrack

The transformer-based TransTrack [7] formulates the object detection and data association in one network, which consists of four main parts, i.e., backbone, encoder, decoder, and prediction heads. The backbone extracted the features of images. The ResNet-50 [47] and FPN [16] are implemented as the backbone. The input of the encoder is the feature map of the image, and the encoder learned the long-range dependency information across the tokens in the feature map. It stacks multi-head self-attention layers and feed-forward networks (FFN) in one block. For the decoder, two parallel modules were applied to detect objects based on the object queries for the current frame and the track queries from the previous frame, respectively, [7]. The two modules have the same architecture, which is built with multi-head self-attention layers, multi-head cross-attention layers and FFN. The inputs for the two modules are the output of the encoder and learnable queries. The difference of the two modules is the query types in the inputs, i.e., the object queries and the track queries. The object queries are a set of learnable parameters, and trained with the other parameters for object detection. The track queries are the reserved output of the module used for detection from previous frame, which are the features of the detected objects in the previous frame. With object queries and track queries, the detection boxes and the track boxes are predicted with two prediction heads, which are the feed forward networks. The data association is conducted by calculating the box IoU scores of the predicted boxes sets. And the Hungarian algorithm is applied for associating each track box with the detection box based on the IoU scores. For the unmatched detection boxes, new trajectories are created. The trajectories of the unmatched track boxes are kept a fixed number of frames since the objects in the trajectories may blocked or disappeared. Although TransTrack accomplish the detection and data association in one network, the slow training convergence and insufficient data association limit its applications.

3.2. Anchor-Based Query

The slow training convergence of the TransTrack is caused by inappropriate design of the learnable queries. The queries are fed into the cross-attention module to learn an attention map with keys, and the content information and spatial information are contained in the queries and keys. It can be described as follows.

$$(C_q + S_q)^T (C_k + S_k) = C_q^T C_k + C_q^T S_k + S_q^T C_k + S_q^T S_k \tag{1}$$

where C_q and C_k are the content part in the queries, S_q and S_k are the spatial part in the keys. The content part of the queries not only need to match the content information in the keys, but also need to match the spatial information in the keys [10]. Moreover, the spatial information in the queries randomly initialized, while the spatial information for the keys are generated with the sinusoidal function function [9]. Therefore, the anchor-based-query is proposed to augment explicit prior spatial knowledge into the queries, which encodes the spatial information of the queries with the sinusoidal function function, to accelerate the training.

The anchor-based-query is implemented in the decoder of the tracking network. As shown in Figure 2, two parts are included in the anchor-based-query (ABQ), i.e., content query and spatial query. Specifically, the content queries and spatial queries for the detection module are the object queries and the anchor boxes, and that for the track module are the track queries and the anchor boxes. The two sets of anchor boxes are generated independently. The content query is same to that in TransTrack, while the spatial query formulates the spatial information in 4D anchors, i.e., (x, y, w, h), which includes both the position and size of the anchor box. The queries are utilized to probe the features of the images, which leads to directly prediction without non-maximum suppression. The 4D anchor boxes are concatenated with the content queries in the channel domain to match the content information and spatial information separately. Therefore, the locations, (x, y) in the 4D anchor boxes, are generated to math the positional embeddings in the keys, which include the locations (x, y). The sizes, (w, h), are utilized for learning the self-attention of the queries with size information, which can adjust the attention maps with scale information. Each element in the anchor boxes is a learnable parameter, which can be learned in the training procedure. As shown in Figure 2, similar ABQ methods are applied in the two modules of the decoder, the detection module and the track module.

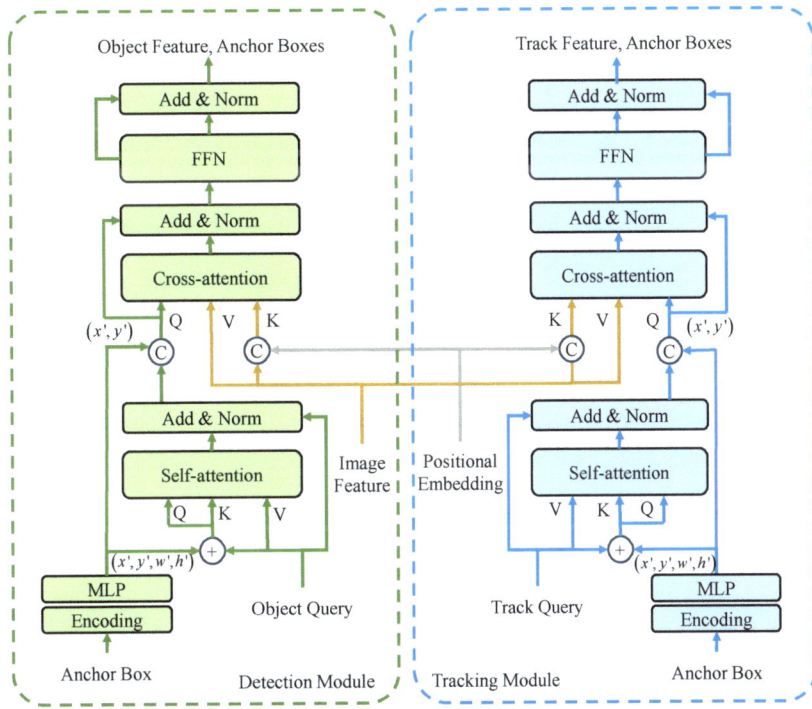

Figure 2. Illustration of the architecture of the decoder based on Anchor-based Query.

For the detection module, self-attention layers and cross-attention layers are used for query updating and feature investigation. The anchor box is defined as $B = (x, y, w, h)$. The positional encoding for each element in the anchor can be calculated as follows.

$$\begin{cases} x'_{2i} = \sin\left(\frac{x}{T_{2i}/D}\right) \\ x'_{2i+1} = \cos\left(\frac{x}{T_{2i+1}/D}\right) \end{cases} \quad (2)$$

where T is a hand-craft temperature as it in [9], $2i$ and $2i + 1$ are the indices, D is the half of the dimension of the content queries. Specifically, the dimension of the content queries is 256 in this work. The positional encoding generates a vector with 128 channels from a float, i.e., the positions. For the self-attention learning, the ABQ can be obtained as follows.

$$B_q = MLP(CAT(x', y', w', h')) \quad (3)$$

where, $MLP(\bullet)$ denotes the multiple layer perceptron, which consists of 2 layers of linear layer with ReLU activation. $CAT(\bullet)$ stands for concatenation operation. The output of the concatenating operation has $2 \times D$ channels, and the MLP operation reduce the dimension of it to D channels so that the spatial queries can be concatenated with the content queries. The queries and keys for the self-attention are the combination of the spatial queries and the content queries as follows.

$$\begin{cases} Q_s = C_q + B_q \\ K_s = C_q + B_q \end{cases} \quad (4)$$

where Q_s and K_s represent the queries and keys for the self-attention learning, respectively, C_q is the content item, which is a set of learnable parameters similar to the object query in TransTrack. The values equal the content item. For the cross-attention, the keys contain a feature map and the corresponding positional embedding, the feature map is outputted from the encoder. The values are the image features same to the keys. They are obtained as follows.

$$\begin{cases} K_c = CAT(F, P) \\ V_c = F \end{cases} \quad (5)$$

where K_c and V_c represent the keys and values for the cross-attention, respectively, F and P stand for the feature map and the corresponding positional embedding. The queries for the cross-attention learning is the union, which consists of content query and the locations in the spatial query, which can be described as follows.

$$Q_c = CAT\left(C'_q, CAT(x', y') \bullet MLP_c(C_q)\right) \quad (6)$$

where C'_q represents the content queries outputted from the self-attention layer, C_q is the content queries same to that in Equation (4), x' and y' are the locations. $MLP_c(\bullet)$ is a multiple layer perceptron, which is utilized to learn a scale reference map on the content query, and $\bullet$ is the element-wise multiplication. The concatenations of the content information and spatial information in the queries and keys enable the cross-attention learning, and decouple the matching of the content information and spatial information.

The content queries and spatial queries for the detection module are sets of learnable parameters, and the queries for the tracking module are the outputs of the detection module. Specifically, the track queries and the ABQ for the tracking module are the object features and anchor boxes outputted from the detection module in the previous frame. The anchor boxes are the positional encoding corresponding to the ABQ.

3.3. Template Matching

To enable data association in JDT methods with feature information, template matching (TM) is augmented in the network. The TBD methods applied additional networks to match the features of the detected objects and the objects in the trajectories. This solution is not suitable for the JDT methods due to the additional network increase the complexity of the JDT methods. Meanwhile, these features are already been extracted in the networks of JDT methods. Therefore, the TM utilizes the features from the tracking networks to compute the feature similarity scores. By comparing the features of the detected objects and the templates corresponding to the trajectories, the ABQ-Track can associate the detection results and trajectories with additional feature information. Moreover, the tracker can maintain the trajectories with long-time memory by integrating the former features into the corresponding templates.

Considering the F^t_{Track} is the feature map outputted from the right branch in the decoder as shown in Figure 2, the superscript t and subscript $Track$ represent the number of the frame and the identity of the trajectory. The template for each trajectory can be computed as follows.

$$T_{Track} = \sum_{t=M-m}^{M} w^t_{Track} F^t_{Track} \quad (7)$$

where T_{Track} represents the template corresponding to the trajectory, $Track$, w^t_{Track} denotes the learnable weights of the features, M and m are the number of current frame and the total number of features, respectively. Taking F_{Object} as the features of detected object, which is the output of the left branch in decoder as shown in Figure 2, each element in the feature similarity score map is obtained with mahalanobis distance as follows.

$$S^{i, Track}_{Feature} = MD\left(F^i_{Object}, T_{Track}\right) \quad (8)$$

where i and $Track$ represents the index of detected objects and the templates, respectively. The $MD(\bullet)$ represents the mahalanobis distance function.

According to Equation (8), the more similar the two features are, the lower $S^{i,Track}_{Feature}$ will be. However, the score map based on spatial information is computed with IoU, which is higher when two boxes share larger overlapping area. Hence, the feature similarity score is further processed as follows.

$$S_{Feature}^{i,Track} = 1 - Norm\left(S_{Feature}^{i,Track}\right) \qquad (9)$$

where $Norm(\bullet)$ is the normalization operation.

3.4. Architecture

The pipeline of the ABQ-Track is illustrated in Figure 3. The two methods proposed in this paper is applied in the decoder and the matching section of the network. For fair evaluation, the rest parts of the network share the same structure to the TransTrack. In the decoder, 6 blocks are implemented for each module, which is described in Section 3.2. The number of the content queries is set to 500, which is based on the experimental results. And the dimension of the content queries is set to 256. The spatial queries for the self-attention and the cross-attention in the decoder have the dimension of 256, and the number is 500. The keys for the self-attention shares the same size with the queries. The keys for the cross-attention are obtained by concatenating the feature outputted from the encoder and the positional embedding. The decoder outputs two sets of feature maps and anchor boxes corresponding to the object queries and the track queries. The feature maps are fed into the two prediction heads to obtain two prediction maps, $m \in R_{500\times 4}$, which indicates the locations of the objects detected with the detection module and the track module. The IoU scores are computed with the locations of the two set of objects. The feature maps are also fed into the TM module to obtain the feature scores.

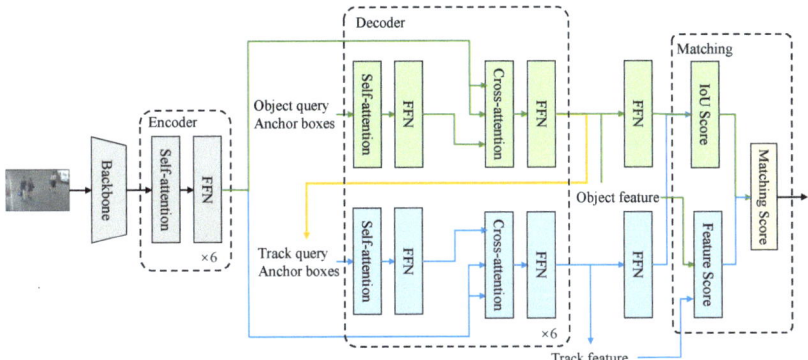

Figure 3. Illustration of the pipeline of the proposed tracker, ABQ-Track.

3.5. Training and Inference

The tracking network is trained with data from CrowdHuman [48] and MOT [13,14]. The data are randomly sampled from a real video clip in the two data sets. Due to the two modules in the decoder perform predictions in the same images, the two modules can be trained with same loss function. Following [37], the loss function of the tracking network can be formulated as follows.

$$L = \lambda_1 L_{cls} + \lambda_2 L_{L_1} + \lambda_3 L_{giou} \qquad (10)$$

where λ_1, λ_2 and λ_3 stand for the weights of the losses. L_{cls} represents the focal loss of classification [17]. L_{L_1} and L_{giou} are the L1 loss and generalized IoU loss [49] for the regression of the coordinates of the predicted boxes.

In the initial image of a given sequence, the ABQ-Track detects only the objects with the parts for detection in the network. Then, the detection and data association are conducted on the rest images of the sequence with the whole tracking network. Two sets of predicted box can be obtained with the features outputted from the decoder based on the object query and track query, respectively. The IoU similarity score map between the two sets of box is computed with the Kuhn-Munkres algorithm. Additionally, the template matching module outputs the similarity score map based on the features of the detected objects and the object within each trajectory. The finally score map can be obtained with the two score maps with weights as follows.

$$S = w_{IoU}S_{IoU} + w_{Feature}S_{Feature} \tag{11}$$

where S_{IoU} and $S_{Feature}$ stand for the score maps of the IoU and feature similarity, w_{IoU} and $w_{Feature}$ represent the corresponding learnable weights. Based on the score map, the traditional Hungarian algorithm [20] is applied for data association. For the unmatched boxes based on the object query, new trajectories are birthed for them. And for the trajectories who have not associated to any boxes based on the object query for 30 consecutive frames, the trajectories are removed. In this way, each box based on the object query has a unique ID.

4. Experiment

To evaluate the proposed methods, experiments have been conducted on commonly applied data sets for multiple object tracking, MOT17 [13] and MOT20 [14]. An ablation study on the effectiveness of proposed methods is presented, and the comparison of the proposed tracker and the other MOT trackers are given to show that the proposed tracker has competitive performance.

4.1. Implementation Details

The network of the ABQ-Track is trained on 2 datasets, CrowdHuman and MOT. The CrowdHuman is utilized to train the network firstly, which is a dataset on scenarios of dense pedestrians. 15,000 training images and 470,000 instances are contained in the CrowdHuman, which provides more categories of the samples. Then, the network is trained with CrowdHuman and MOT. The training data in MOT is split into two subsets for training and validation, respectively. Regular data augmentations are applied on the training data, which include random horizontal, random crop and scale augmentation. The trained tracking network is tested locally with the testing data in MOT dataset, and the results are uploaded to the given server for evaluation. The evaluation metrics in the experiments are MOTA, IDF1, FP, FN, IDS, et al., which follow that in [13]. MOTA is the metric specially designed for MOT task, which is calculated with false positive (FP), false negative (FN), ID switches (IDS) and ground truth (GT).

The backbone in the proposed ABQ-Track is ResNet-50, which is same with the TransTrack. The encoder and decoder have 6 blocks. In each block of the encoder, a self-attention module and a FFN is applied. The structure of the decoder blocks is shown in Figure 2. Each of the two prediction heads have 3 layers of perceptron with ReLU activation function and a linear projection layer. The parameters of the backbone are initialized with that learned on ImageNet [50], and the rest parameters are initialized with Xaxiver-init [51]. There are two steps of training of the tracking network, and the optimizer is AdamW. Firstly, the tracking network is trained on the CrowdHuman for 50 epochs, and the batch size is set to 16. The learning rate is set to 2×10^{-4}, and it drops to 2×10^{-5} after 30th epoch. Secondly, the tracking network is fine-tuned for 40 epochs, including CrowdHuman and the splits of MOT. The learning rate and the batch size for the fine-tuning are set to 2×10^{-5} and 16. Commonly used data augmentations, i.e., random horizontal, random crop, scale augmentation, are applied for the two training steps. The weights of the losses are set to $\lambda_1 = 2; \lambda_2 = 5; \lambda_3 = 2$.

4.2. MOT Challenge Test Results and Discussion

4.2.1. MOT17

The upper section of Table 1 reports the evaluation results of ABQ-Track and 10 other trackers on the MOT17 test dataset. ABQ-Track, which is pre-trained with the CrowdHuman dataset, achieved the best score in MOTA, FP and FN. In terms of the other metrics, competitive performance is achieved. The better performances of SOTMOT, CSTrack and TransCenter in IDF1 are achieved with larger pre-train data set. ABQ-Track chooses the CrowdHuman for fair comparison with TransTrack. Comparing to TransTrack, ABQ-Track achieves better performance in MOTA, IDF1, FP, MT, ML and IDS. The better scores in MT and ML can be attributed to the implementation of the ABQ, which improves the detection performance of the tracking network. The utilization of TM reduces the IDS, for the feature mathcing stablizing the trajectories. The ABQ and TM work together to improve the performance in MOTA. Notably, the network of TransTrack is needed training for 150 epochs, while that of ABQ-Track is trained in only 50 epochs.

Table 1. Results on the MOT17 and MOT20 Test Sets: The upper section presents results on private detection in the MOT17 dataset, while the lower section displays the detection outcomes on the MOT20 dataset.The upward (downward) arrows in the table indicate that the larger (smaller) the parameter, the better the corresponding performance.

Method	MOTA↑	IDF1↑	FP↓	FN↓	MT↑	ML↓	IDS↓
			MOT17				
UMA [52]	53.1	54.4	22,893	239,534	21.5	31.8	2251
TubeTK [53]	63.0	58.6	27,060	177,483	31.2	19.9	4137
CenterTrack [32]	67.8	64.7	18,489	160,332	34.6	24.6	3039
QuasiDense [2]	68.7	66.3	26,589	146,643	43.8	17.22	3378
TraDeS [54]	69.1	63.9	20,892	150,060	36.4	21.5	3555
SOTMOT [55]	71.0	71.9	39,537	118,983	42.7	15.3	5184
TransCenter [56]	72.5	58.1	25,722	114,310	64.7	12.2	2332
FairMOT [57]	73.7	72.3	27,507	117,477	43.2	17.3	3303
TransTrack [7]	74.5	63.9	28,323	112,137	46.8	11.3	3663
CSTrack [58]	74.9	72.6	23,847	114,303	41.5	17.5	3567
ABQ-Track (ours)	75.9	65.4	16,977	115,667	52.8	2.8	3135
			MOT20				
FairMOT [57]	61.8	67.3	103,440	88,901	66.3	8.5	5243
TransTrack [7]	64.5	59.2	28,566	151,377	49.1	13.6	3565
CorrTracker [59]	65.2	69.1	79,429	95,855	66.4	8.9	5183
CSTrack [58]	66.6	68.6	25,404	144,358	50.4	15.5	3196
ABQ-Track (ours)	66.3	60.3	20,179	149,536	38.0	45.1	3383

4.2.2. MOT20

The lower section of Table 1 presents the performance of ABQ-Track and the other 5 trackers on the MOT20 test dataset. MOT20 consists of 4 training sequences and 4 testing sequences, which are obtained in more challenging tracking scenarios. Specifically, the test data in MOT20 contain extremely dense scenarios. As shown in Table 1, ABQ-Track achieves competitive performance in the comparison with the other 5 trackers. The trackers, i.e., CSTrack, CorrTrack, SOTMOT, surpass the performance of ABQ-Track for larger training dataset. However, the IDS of ABQ-Track is lower for the utilization of the TM. Compared to TransTrack, ABQ-Track achieves better performance in all the metrics. which reveals that the proposed modules improves the performance of the JDT trackers in dense multiple object tracking scenaros.

4.3. Ablation Study

4.3.1. Anchor-Based-Query

In this section, the ABQ-Track networks trained for different numbers of epochs are evaluated on the validation data in MOT17 since that the motivation of utilizing the ABQ is to reduce the training epochs of JDT methods. Table 2 reports the performance results of the trackers in this experiment. All models are trained with the same data. TransTrack, ABQ-10, ABQ-25, and ABQ-50 are trained in 150 epochs, 10 epochs, 25 epochs and 50 epochs, respectively. The tracker with ABQ achieves convergence in 50 epochs, and further training brings no performance improvement. Comparing to TransTrack, the tracker with ABQ has faster training procedure and better performance. As the queries in the decoder works as pooling feature from a feature map. The content queries represents the semantic information, while the spatial queries constrain the pooling feature around the content queries. The Anchor-based-Query leads the tracker to focus on the local area around the corresponding object. This is the main reason for the better performance and faster convergence than that of TransTrack.

Table 2. Comparison of the tracker with ABQ trained in different epochs. The experiment is conducted on the validation data in MOT17. The IDs represents the percentage of the number of the ID switches to all the identities. The upward (downward) arrows in the table indicate that the larger (smaller) the parameter, the better the corresponding performance.

Architecture	MOTA↑	FP↓	FN↓	IDF1↑	IDs↓
TransTrack	65.0%	4.3%	30.3%	-	0.4%
ABQ-10	37.8%	33.1%	68.5%	44.8%	17.1%
ABQ-25	54.8%	19.3%	45.4%	56.8%	13.7%
ABQ-50	66.1%	3.9%	28.6%	66.7%	0.6%
ABQ-75	65.8%	4.1%	29.0%	60.2%	10.5%

Moreover, the performances of the trackers with different ABQ designs are reported in Table 3. The Location represents the generating ABQ locations. The Adding stands for the augmenting the ABQ into the object queries and track queries with summation. The ABQ in the table represents the methods described in Section 3.1. All trackers are trained for 50 epochs, and with the same training data. The None tracker achieves lowest performance scores since it needs more epochs to convergence. The Adding tracker have similar performance as the None tracker due to summation mixes the content queries and anchor-box queries for the attention learning. Comparing the Location and ABQ, adding scale information in the ABQ improves the performance of the trackers.

Table 3. Comparison of the trackers with different ABQ designs. The upward (downward) arrows in the table indicate that the larger (smaller) the parameter, the better the corresponding performance.

Trackers	MOTA↑	FP↓	FN↓	IDF1↑	IDs↓
None	54.5%	16.0%	33.7%	57.9%	13.9%
Location	64.8%	4.3%	29.6%	65.2%	0.6%
Adding	55.3%	15.8%	33.8%	58.2%	13.1%
ABQ	66.1%	3.9%	28.6%	66.7%	0.6%

Additionally, the comparison on performances of the trackers with different numbers of the ABQ is conducted due to the quantity of queries has obvious affect on the performance of the transformer-based detection methods. The experimental results are reported in Table 4. Comparing the trackers with 300 anchor-based-queries and 500 anchor-based-queries, more anchor-based-queries improve the performance of the tracker. And further adding more anchor-based queries does not lead to obvious performance improvement. This is because too many anchor-based queries leads to excessive fragmentation of the feature map, which results in confusion between the localization targets in dense scenarios. Based on this experiment, the number of the anchor-based-queries in ABQ-Track is set to 500.

Table 4. The experimental results on the comparison of different anchor-based-queries settings. The upward (downward) arrows in the table indicate that the larger (smaller) the parameter, the better the corresponding performance.

num_Query	MOTA↑	FP↓	FN↓	IDF1↑	IDs↓
300	66.7%	6.6%	25.3%	63.7%	1.4%
500	68.6%	4.3%	26.2%	70.1%	0.9%
900	68.4%	5.5%	24.9%	68.3%	1.1%

4.3.2. Template Matching

The classical JDT methods, TransTrack, focus on associating the detected objects and trajectories based on the spatial score, i.e., IoU score. Id switches often occur when objects are occluded during tracking or objects with similar appearances appear. Preserving historical feature information can benefit for maintaining the stable trajectories. Hence, to avoid identity switches and maintain stable trajectories, the TM module is designed for TransTrack. The TM module bring historical feature information, add feature-based score into the data association method in the tracker. In the section, the performance comparison is made between the trackers with or without the TM module. Meanwhile, the amount of past feature information saved in the template is evaluated. The results are reported in Table 5.

Table 5. Performance comparison for TM modules. The upward (downward) arrows in the table indicate that the larger (smaller) the parameter, the better the corresponding performance.

Trackers	MOTA↑	FP↓	FN↓	IDF1↑	IDs↓
MOT-None	66.1%	3.9%	28.6%	66.7%	0.6%
TM-Input	66.4%	5.5%	25.5%	67.1%	1.2%
TM-5	66.6%	5.5%	25.5%	67.7%	1.3%
TM-10	66.9%	5.4%	25.1%	67.9%	1.2%
TM-20	67.6%	4.7%	24.9%	68.9%	1.3%
TM-30	68.6%	4.3%	26.2%	70.1%	0.9%

In Table 5, the None represents the ABQ-Track without the TM module. TM-Input stands for the ABQ-Track with the TM module, and only one feature is saved to obtain the template. TM-5, TM-10, TM-20, TM-30 stand for the TM utilizes 5, 10, 20 and 30 features from the past to obtain the template, respectively. As shown in Table 5, trackers apply the TM modules outperform the tracker without the TM modules, which verifies the effectiveness of the TM module. Moreover, TM-30 achieves the best performance among the trackers with the TM module. Hence, the ABQ-track in Table 1 utilizes 30 features from the past to obtain the template.

4.3.3. Visualization of Experimental Results

To better illustrate the enhancement in MOT tracking performance with the ABQ and TM, the visualization of the tracking results of the trackers in one sequence is presented in Figure 4. These results are all derived from the MOT17-02 validation video sequence. The first row represents the tracking results of TransTrack, the second row depicts the results of the tracker applying the ABQ, and the third row illustrates the tracking result of the ABQ-Track.

As shown in Figure 4, the tracker with Anchor-based query detected the object which is missed by the TransTrack, and certain ID switch in the results of TransTrack does not occur in the results of the tracker with Anchor-based query. This reveals that the ABQ improves the detection performance and reduces the ID switches of the tracker. Comparing the second and third row, the ID switches is further reduced, which verifies the effectiveness of the TM.

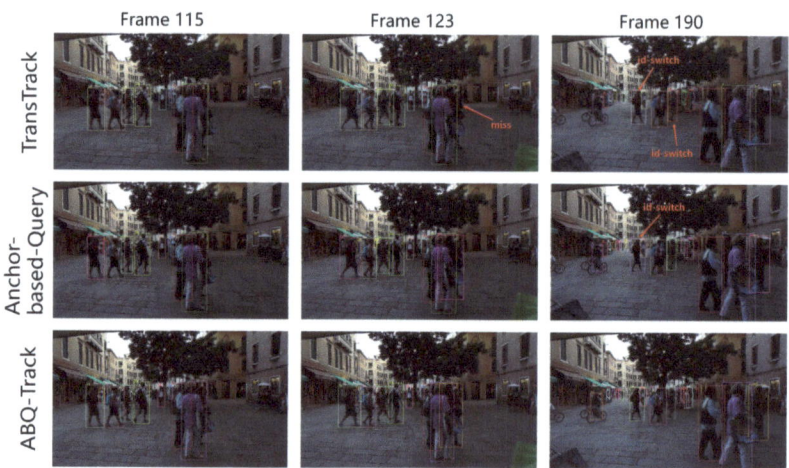

Figure 4. Visualization of tracking results of TransTrack, tracker with ABQ and ABQ-Track.

5. Conclusions

In this paper, a JDT method based on a transformer, ABQ-Track, is proposed. First, the explicit spatial knowledge is augmented in the query of the decoder with a four-dimensional anchor box (ABQ). The ABQ increases the performance of the tracker, and leads to faster training convergence. Additionally, a new template matching module (TM) is proposed to enable the JDT methods to associate the trajectories and detection results based on historical features. Based on the two modules, the ABQ-Track is built by modifying the classical JDT method, TransTrack, and achieves better performance than TransTrack on MOT 17 and MOT 20, 75.9% and 66.3% in MOTA, respectively. Specifically, the ABQ-Track achieves the better performance after being trained for 50 epochs, while the TransTrack needs to be trained for 150 epochs.

Author Contributions: Conceptualization, Q.W.; Software, C.L.; Validation, C.L.; Formal analysis, Q.W.; Investigation, Q.W.; Writing—original draft, L.G.; Writing—review & editing, C.L. and L.G.; Visualization, G.H.; Supervision, L.G. All authors have read and agreed to the published version of the manuscript.

Funding: This research was funded by project CEIEC-2022-ZM02-0247.

Institutional Review Board Statement: Not applicable.

Informed Consent Statement: Not applicable.

Data Availability Statement: The public data used in this work are listed here: CrowdHuman www.crowdhuman.org/ (accessed on 28 November 2023), MOT17 https://motchallenge.net/ (accessed on 28 November 2023), MOT20 https://motchallenge.net/ (accessed on 28 November 2023).

Conflicts of Interest: The authors declare no conflict of interest.

References

1. Cheng, C.C.; Qiu, M.X.; Chiang, C.K.; Lai, S.H. ReST: A Reconfigurable Spatial-Temporal Graph Model for Multi-Camera Multi-Object Tracking. In Proceedings of the IEEE/CVF International Conference on Computer Vision, Paris, France, 2–6 October 2023; pp. 10051–10060.
2. Pang, J.; Qiu, L.; Li, X.; Chen, H.; Li, Q.; Darrell, T.; Yu, F. Quasi-dense similarity learning for multiple object tracking. In Proceedings of the IEEE/CVF Conference on Computer Vision and Pattern Recognition, Online, 19–25 June 2021; pp. 164–173.

3. Zhang, Y.; Sun, P.; Jiang, Y.; Yu, D.; Weng, F.; Yuan, Z.; Luo, P.; Liu, W.; Wang, X. Bytetrack: Multi-object tracking by associating every detection box. In Proceedings of the European Conference on Computer Vision, Tel-Aviv, Israel, 23–27 October 2022; pp. 1–21.
4. Woo, S.; Park, K.; Oh, S.W.; Kweon, I.S.; Lee, J.Y. Tracking by Associating Clips. In Proceedings of the European Conference on Computer Vision, Tel-Aviv, Israel, 23–27 October 2022; pp. 129–145.
5. Zhou, X.; Yin, T.; Koltun, V.; Krähenbühl, P. Global tracking transformers. In Proceedings of the IEEE/CVF Conference on Computer Vision and Pattern Recognition, New Orleans, LA, USA, 19–24 June 2022; pp. 8771–8780.
6. Kieritz, H.; Hubner, W.; Arens, M. Joint detection and online multi-object tracking. In Proceedings of the IEEE Conference on Computer Vision and Pattern Recognition Workshops, Salt Lake City, UT, USA, 18–22 June 2018; pp. 1459–1467.
7. Sun, P.; Cao, J.; Jiang, Y.; Zhang, R.; Xie, E.; Yuan, Z.; Wang, C.; Luo, P. Transtrack: Multiple object tracking with transformer. *arXiv* **2020**, arXiv:2012.15460.
8. Meinhardt, T.; Kirillov, A.; Leal-Taixe, L.; Feichtenhofer, C. Trackformer: Multi-object tracking with transformers. In Proceedings of the IEEE/CVF Conference on Computer Vision and Pattern Recognition, New Orleans, LA, USA, 19–24 June 2022; pp. 8844–8854.
9. Liu, S.; Li, F.; Zhang, H.; Yang, X.; Qi, X.; Su, H.; Zhu, J.; Zhang, L. DAB-DETR: Dynamic Anchor Boxes are Better Queries for DETR. In Proceedings of the International Conference on Learning Representations, Online, 25–29 April 2022.
10. Meng, D.; Chen, X.; Fan, Z.; Zeng, G.; Li, H.; Yuan, Y.; Sun, L.; Wang, J. Conditional detr for fast training convergence. In Proceedings of the IEEE/CVF International Conference on Computer Vision, Montreal, QC, Canada, 10–17 October 2021; pp. 3651–3660.
11. Cai, J.; Xu, M.; Li, W.; Xiong, Y.; Xia, W.; Tu, Z.; Soatto, S. Memot: Multi-object tracking with memory. In Proceedings of the IEEE/CVF Conference on Computer Vision and Pattern Recognition, New Orleans, LA, USA, 19–24 June 2022; pp. 8090–8100.
12. Li, S.; Kong, Y.; Rezatofighi, H. Learning of global objective for network flow in multi-object tracking. In Proceedings of the IEEE/CVF Conference on Computer Vision and Pattern Recognition, New Orleans, LA, USA, 19–24 June 2022; pp. 8855–8865.
13. Milan, A.; Leal-Taixé, L.; Reid, I.; Roth, S.; Schindler, K. MOT16: A benchmark for multi-object tracking. *arXiv* **2016**, arXiv:1603.00831.
14. Dendorfer, P.; Rezatofighi, H.; Milan, A.; Shi, J.; Cremers, D.; Reid, I.; Roth, S.; Schindler, K.; Leal-Taixé, L. Mot20: A benchmark for multi object tracking in crowded scenes. *arXiv* **2020**, arXiv:2003.09003.
15. Dave, A.; Khurana, T.; Tokmakov, P.; Schmid, C.; Ramanan, D. Tao: A large-scale benchmark for tracking any object. In Proceedings of the European Conference on Computer Vision, Online, 23–28 August 2020; pp. 436–454.
16. Lin, T.Y.; Dollár, P.; Girshick, R.; He, K.; Hariharan, B.; Belongie, S. Feature pyramid networks for object detection. In Proceedings of the IEEE Conference on Computer Vision and Pattern Recognition, Hawaii, HI, USA, 21–26 July 2017; pp. 2117–2125.
17. Lin, T.Y.; Goyal, P.; Girshick, R.; He, K.; Dollár, P. Focal loss for dense object detection. In Proceedings of the IEEE International Conference on Computer Vision, Venice, Italy, 22–29 October 2017; pp. 2980–2988.
18. Ren, S.; He, K.; Girshick, R.; Sun, J. Faster r-cnn: Towards real-time object detection with region proposal networks. In Proceedings of the 28th International Conference on Neural Information Processing Systems, Plaza Hotel Bali, Indonesia, 8–12 December 2021; pp. 91–99.
19. Redmon, J.; Farhadi, A. Yolov3: An incremental improvement. *arXiv* **2018**, arXiv:1804.02767.
20. Bewley, A.; Ge, Z.; Ott, L.; Ramos, F.; Upcroft, B. Simple online and realtime tracking. In Proceedings of the 2016 IEEE International Conference on Image Processing, Phoenix, AZ, USA, 25–28 September 2016; pp. 3464–3468.
21. Wojke, N.; Bewley, A.; Paulus, D. Simple online and realtime tracking with a deep association metric. In Proceedings of the 2017 IEEE International Conference on Image Processing, Beijing, China, 17–20 September 2017; pp. 3645–3649.
22. Wang, Z.; Zheng, L.; Liu, Y.; Li, Y.; Wang, S. Towards real-time multi-object tracking. In Proceedings of the European Conference on Computer Vision, Online, 23–28 August 2020; pp. 107–122.
23. Du, Y.; Zhao, Z.; Song, Y.; Zhao, Y.; Su, F.; Gong, T.; Meng, H. Strongsort: Make deepsort great again. *IEEE Trans. Multimed.* **2023**, *25*, 8725–8737. [CrossRef]
24. Du, Y.; Wan, J.; Zhao, Y.; Zhang, B.; Tong, Z.; Dong, J. Giaotracker: A comprehensive framework for mcmot with global information and optimizing strategies in visdrone 2021. In Proceedings of the IEEE/CVF International Conference on Computer Vision, Montreal, QC, Canada, 10–17 October 2021; pp. 2809–2819.
25. Han, S.; Huang, P.; Wang, H.; Yu, E.; Liu, D.; Pan, X. Mat: Motion-aware multi-object tracking. *Neurocomputing* **2022**, *476*, 75–86. [CrossRef]
26. Stadler, D.; Beyerer, J. Modelling ambiguous assignments for multi-person tracking in crowds. In Proceedings of the IEEE/CVF Winter Conference on Applications of Computer Vision, Waikoloa, HI, USA, 4–8 January 2022; pp. 133–142.
27. Aharon, N.; Orfaig, R.; Bobrovsky, B.Z. BoT-SORT: Robust associations multi-pedestrian tracking. *arXiv* **2022**, arXiv:2206.14651.
28. Leal-Taixé, L.; Canton-Ferrer, C.; Schindler, K. Learning by tracking: Siamese CNN for robust target association. In Proceedings of the IEEE Conference on Computer Vision and Pattern Recognition Workshops, Las Vegas, NV, USA, 26 June–1 July 2016; pp. 33–40.
29. Ristani, E.; Tomasi, C. Features for multi-target multi-camera tracking and re-identification. In Proceedings of the IEEE Conference on Computer Vision and Pattern Recognition, Istanbul, Turkey, 30–31 January 2018; pp. 6036–6046.
30. Feichtenhofer, C.; Pinz, A.; Zisserman, A. Detect to track and track to detect. In Proceedings of the IEEE International Conference on Computer Vision, Venice, Italy, 22–29 October 2017; pp. 3038–3046.

31. Bergmann, P.; Meinhardt, T.; Leal-Taixe, L. Tracking without bells and whistles. In Proceedings of the IEEE/CVF International Conference on Computer Vision, Seoul, Republic of Korea, 27 October–2 November 2019; pp. 941–951.
32. Zhou, X.; Koltun, V.; Krähenbühl, P. Tracking objects as points. In Proceedings of the European Conference on Computer Vision, Online, 23–28 August 2020; pp. 474–490.
33. Zhang, Y.; Sheng, H.; Wu, Y.; Wang, S.; Ke, W.; Xiong, Z. Multiplex labeling graph for near-online tracking in crowded scenes. *IEEE Internet Things J.* **2020**, *7*, 7892–7902. [CrossRef]
34. Vaswani, A.; Shazeer, N.; Parmar, N.; Uszkoreit, J.; Jones, L.; Gomez, A.N.; Kaiser, Ł.; Polosukhin, I. Attention is all you need. In Proceedings of the Advances in Neural Information Processing Systems, Long Beach, CA, USA, 4–9 December 2017; pp. 5998–6008.
35. Wu, H.; Xiao, B.; Codella, N.; Liu, M.; Dai, X.; Yuan, L.; Zhang, L. Cvt: Introducing convolutions to vision transformers. In Proceedings of the IEEE/CVF International Conference on Computer Vision, Montreal, QC, Canada, 10–17 October 2021; pp. 22–31.
36. Liu, Z.; Lin, Y.; Cao, Y.; Hu, H.; Wei, Y.; Zhang, Z.; Lin, S.; Guo, B. Swin transformer: Hierarchical vision transformer using shifted windows. In Proceedings of the IEEE/CVF International Conference on Computer Vision, Montreal, QC, Canada, 10–17 October 2021; pp. 10012–10022.
37. Carion, N.; Massa, F.; Synnaeve, G.; Usunier, N.; Kirillov, A.; Zagoruyko, S. End-to-end object detection with transformers. In Proceedings of the European Conference on Computer Vision, Online, 23–28 August 2020; pp. 213–229.
38. Li, Y.; Mao, H.; Girshick, R.; He, K. Exploring plain vision transformer backbones for object detection. In Proceedings of the European Conference on Computer Vision, Tel-Aviv, Israel, 23–27 October 2022; pp. 280–296.
39. Wang, Y.; Xu, Z.; Wang, X.; Shen, C.; Cheng, B.; Shen, H.; Xia, H. End-to-end video instance segmentation with transformers. In Proceedings of the IEEE/CVF Conference on Computer Vision and Pattern Recognition, Online, 19–25 June 2021; pp. 8741–8750.
40. Dosovitskiy, A.; Beyer, L.; Kolesnikov, A.; Weissenborn, D.; Zhai, X.; Unterthiner, T.; Dehghani, M.; Minderer, M.; Heigold, G.; Gelly, S.; et al. An image is worth 16x16 words: Transformers for image recognition at scale. *arXiv* **2020**, arXiv:2010.11929.
41. Wang, W.; Xie, E.; Li, X.; Fan, D.P.; Song, K.; Liang, D.; Lu, T.; Luo, P.; Shao, L. Pyramid vision transformer: A versatile backbone for dense prediction without convolutions. In Proceedings of the IEEE/CVF international conference on computer vision, Montreal, QC, Canada, 10–17 October 2021; pp. 568–578.
42. Chen, Z.; Zhu, Y.; Zhao, C.; Hu, G.; Zeng, W.; Wang, J.; Tang, M. Dpt: Deformable patch-based transformer for visual recognition. In Proceedings of the 29th ACM International Conference on Multimedia, Chengdu, China, 20–24 October 2021; pp. 2899–2907.
43. Dong, X.; Bao, J.; Chen, D.; Zhang, W.; Yu, N.; Yuan, L.; Chen, D.; Guo, B. Cswin transformer: A general vision transformer backbone with cross-shaped windows. In Proceedings of the IEEE/CVF Conference on Computer Vision and Pattern Recognition, New Orleans, LA, USA, 19–24 June 2022; pp. 12124–12134.
44. Tu, Z.; Talebi, H.; Zhang, H.; Yang, F.; Milanfar, P.; Bovik, A.; Li, Y. Maxvit: Multi-axis vision transformer. In Proceedings of the European Conference on Computer Vision, Tel-Aviv, Israel, 23–27 October 2022; pp. 459–479.
45. Zhu, X.; Su, W.; Lu, L.; Li, B.; Wang, X.; Dai, J. Deformable detr: Deformable transformers for end-to-end object detection. In Proceedings of the 9th International Conference on Learning Representations, Virtual Event, Austria, 3–7 May 2021; pp. 1–11.
46. Dai, X.; Chen, Y.; Yang, J.; Zhang, P.; Yuan, L.; Zhang, L. Dynamic detr: End-to-end object detection with dynamic attention. In Proceedings of the IEEE/CVF International Conference on Computer Vision, Montreal, QC, Canada, 10–17 October 2021; pp. 2988–2997.
47. He, K.; Zhang, X.; Ren, S.; Sun, J. Deep residual learning for image recognition. In Proceedings of the IEEE Conference on Computer Vision and Pattern Recognition, Las Vegas, NV, USA, 26 June–1 July 2016; pp. 770–778.
48. Shao, S.; Zhao, Z.; Li, B.; Xiao, T.; Yu, G.; Zhang, X.; Sun, J. Crowdhuman: A benchmark for detecting human in a crowd. *arXiv* **2018**, arXiv:1805.00123.
49. Rezatofighi, H.; Tsoi, N.; Gwak, J.; Sadeghian, A.; Reid, I.; Savarese, S. Generalized intersection over union: A metric and a loss for bounding box regression. In Proceedings of the IEEE/CVF Conference on Computer Vision and Pattern Recognition, Long Beach, CA, USA, 16–20 June 2019; pp. 658–666.
50. Deng, J.; Dong, W.; Socher, R.; Li, L.J.; Li, K.; Fei-Fei, L. Imagenet: A large-scale hierarchical image database. In Proceedings of the 2009 IEEE Conference on Computer Vision and Pattern Recognition, Miami, FL, USA, 20–25 June 2009; pp. 248–255.
51. Glorot, X.; Bengio, Y. Understanding the difficulty of training deep feedforward neural networks. In Proceedings of the Thirteenth International Conference on Artificial Intelligence and Statistics, Sardinia, Italy, 13–15 May 2010; pp. 249–256.
52. Yin, J.; Wang, W.; Meng, Q.; Yang, R.; Shen, J. A unified object motion and affinity model for online multi-object tracking. In Proceedings of the IEEE/CVF Conference on Computer Vision and Pattern Recognition, Seattle, WA, USA, 14–19 June 2020; pp. 6768–6777.
53. Pang, B.; Li, Y.; Zhang, Y.; Li, M.; Lu, C. Tubetk: Adopting tubes to track multi-object in a one-step training model. In Proceedings of the IEEE/CVF Conference on Computer Vision and Pattern Recognition, Seattle, WA, USA, 14–19 June 2020; pp. 6308–6318.
54. Wu, J.; Cao, J.; Song, L.; Wang, Y.; Yang, M.; Yuan, J. Track to detect and segment: An online multi-object tracker. In Proceedings of the IEEE/CVF Conference on Computer Vision and Pattern Recognition, Online, 19–25 June 2021; pp. 12352–12361.
55. Zheng, L.; Tang, M.; Chen, Y.; Zhu, G.; Wang, J.; Lu, H. Improving multiple object tracking with single object tracking. In Proceedings of the IEEE/CVF Conference on Computer Vision and Pattern Recognition, Online, 19–25 June 2021; pp. 2453–2462.

56. Xu, Y.; Ban, Y.; Delorme, G.; Gan, C.; Rus, D.; Alameda-Pineda, X. Transcenter: Transformers with dense queries for multiple-object tracking. *IEEE Trans. Pattern Anal. Mach. Intell.* **2021**, *15*, 7820–7835.
57. Zhang, Y.; Wang, C.; Wang, X.; Zeng, W.; Liu, W. A simple baseline for multi-object tracking. *Int. J. Comput. Vis.* **2021**, *7*, 1–19.
58. Liang, C.; Zhang, Z.; Zhou, X.; Li, B.; Zhu, S.; Hu, W. Rethinking the competition between detection and reid in multiobject tracking. *IEEE Trans. Image Process.* **2022**, *31*, 3182–3196. [CrossRef] [PubMed]
59. Wang, Q.; Zheng, Y.; Pan, P.; Xu, Y. Multiple object tracking with correlation learning. In Proceedings of the IEEE/CVF Conference on Computer Vision and Pattern Recognition, Online, 19–25 June 2021; pp. 3876–3886.

Disclaimer/Publisher's Note: The statements, opinions and data contained in all publications are solely those of the individual author(s) and contributor(s) and not of MDPI and/or the editor(s). MDPI and/or the editor(s) disclaim responsibility for any injury to people or property resulting from any ideas, methods, instructions or products referred to in the content.

Article

Multi-Frame Content-Aware Mapping Network for Standard-Dynamic-Range to High-Dynamic-Range Television Artifact Removal

Zheng Wang and Gang He *

School of Telecommunications Engineering, Xidian University, Xi'an 710071, China; jackwu0630@gmail.com
* Correspondence: ghe@xidian.edu.cn

Abstract: Recently, advancements in image sensor technology have paved the way for the proliferation of high-dynamic-range television (HDRTV). Consequently, there has been a surge in demand for the conversion of standard-dynamic-range television (SDRTV) to HDRTV, especially due to the dearth of native HDRTV content. However, since SDRTV often comes with video encoding artifacts, SDRTV to HDRTV conversion often amplifies these encoding artifacts, thereby reducing the visual quality of the output video. To solve this problem, this paper proposes a multi-frame content-aware mapping network (MCMN), aiming to improve the performance of conversion from low-quality SDRTV to high-quality HDRTV. Specifically, we utilize the temporal spatial characteristics of videos to design a content-aware temporal spatial alignment module for the initial alignment of video features. In the feature prior extraction stage, we innovatively propose a hybrid prior extraction module, including cross-temporal priors, local spatial priors, and global spatial prior extraction. Finally, we design a temporal spatial transformation module to generate an improved tone mapping result. From time to space, from local to global, our method makes full use of multi-frame information to perform inverse tone mapping of single-frame images, while it is also able to better repair coding artifacts.

Keywords: standard dynamic range (SDR); high dynamic range (HDR); video coding; artifact removal

Citation: Wang, Z.; He, G. Multi-Frame Content-Aware Mapping Network for Standard-Dynamic-Range to High-Dynamic-Range Television Artifact Removal. *Sensors* **2024**, *24*, 299. https://doi.org/10.3390/s24010299

Academic Editor: Yun Zhang

Received: 20 November 2023
Revised: 22 December 2023
Accepted: 29 December 2023
Published: 4 January 2024

Copyright: © 2024 by the authors. Licensee MDPI, Basel, Switzerland. This article is an open access article distributed under the terms and conditions of the Creative Commons Attribution (CC BY) license (https://creativecommons.org/licenses/by/4.0/).

1. Introduction

Over the past few decades, television production technology has seen rapid advancements. From standard definition (SD) to high definition (HD) and then to ultra-high definition (UHD or 4K, and even 8K), the progress in television and video technology is evident. Central to these advancements is the role of improved sensor technology, which has enabled the capture of richer details and more accurate color reproduction. Similarly, advancements in color gamut and dynamic range have transitioned from the BT.709 standard to the BT.2020 standard, largely due to the capabilities of these advanced sensors. Concurrently, standard dynamic range (SDR) technology has gradually evolved into high dynamic range (HDR) technology, offering audiences a more realistic, dynamic, and color-rich visual experience.

The incorporation of enhanced sensors in modern cameras has been pivotal in HDR technology. They have been implemented in many modern display devices and televisions, providing users with a higher contrast and a richer color representation. However, despite the advancements in sensor technology, the availability of HDR video content remains relatively low. Most of the existing video resources are still provided in SDR format, which to some extent limits consumers' opportunities to experience HDR display technology. This phenomenon might be attributed to various factors, including but not limited to the cost of producing HDR content and technical requirements. Therefore, while the proliferation of HDR display technology offers the potential for elevating video content quality, the production and distribution of HDR video content still face some challenges.

Upon capturing light, a camera equipped with state-of-the-art sensors performs several processes before displaying the video image on a monitor. Initially, the light signal is transformed into a digital signal via a CMOS sensor. Following this, tone mapping [1] is employed to convert the high-dynamic digital signal captured by the sensor into a low-dynamic signal. Subsequently, gamut mapping adjusts the image color to match the target color gamut. The linear signal is then altered to a nonlinear signal through the optical electronic transfer function (OETF) [2,3]. This digital signal is then quantized and subjected to arithmetic coding [4], making it ready for encoding and decoding through a codec [5,6]. After decoding, the nonlinear signal is reverted to a linear digital signal via an electronic optical transfer function (EOTF), which is then converted into an optical signal for playback on the monitor. The primary distinction between SDR and HDR lies in the utilization of different EOTFs and OETFs, which are crucial in rendering image brightness and color.

HDRTVs, compared to SDRTVs, offer notable advancements in visual perception, among other aspects. However, a significant portion of current video resources remain in SDR format, a historical issue rooted in the capabilities of earlier video recording hardware and sensors that stored videos in the SDRTV format. Given this, devising a solution for converting SDRTV to HDRTV, especially leveraging the data from advanced sensors, is valuable. In this discourse, the process of converting SDRTV to HDRTV is represented as SDRTV-to-HDRTV, aligning with the method mentioned in [7]. SDR-to-HDRTV denotes the conversion from an SDR television image to an HDR television image, where HDR television images, especially those captured with modern sensors, can be displayed on devices through tone mapping.

Earlier approaches [8–10] combined super-resolution techniques with SDRTV-to-HDRTV conversion, endeavoring to establish a pipeline to transition from low-resolution SDR video to high-resolution HDR video. In a different vein, HDRTVNET [7] introduced a multi-stage mechanism to achieve SDRTV-to-HDRTV conversion, employing global tone mapping, local image enhancement, and image generation. Similarly, the HDCFM framework [11] involves hierarchical dynamic context feature mapping to facilitate the learning of the mapping function from the SDR frame to the HDR frame.

As mentioned above, historical technical constraints and copyright issues have resulted in a vast quantity of current SDRTV videos lacking nearly lossless versions, leaving only relatively low-quality SDRTV versions available. The practical application of the SDRTV-to-HDRTV method necessitates the conversion of low-quality (LQ) SDRTV to high-quality (HQ) HDRTV. Concurrently, prior research [12,13] discovered that the traditional technique of transitioning from LQ SDRTV to HQ HDRTV tends to magnify the coding artifacts.

In particular, as illustrated in the left figure of Figure 1, applying inverse tone mapping to LQ SDRTV significantly amplifies blocking artifacts. Likewise, the right part reveals how banding artifacts are intensified due to the oversight of encoding compression during the conversion process. These observations confirm that the process of converting SDRTV to HDRTV often exacerbates the encoding artifacts inherent in SDRTV, which in turn diminishes the visual quality of the resultant video. The challenges posed by encoding artifacts are a crucial consideration in developing and refining methods for SDRTV-to-HDRTV conversion in order to attain a superior visual output in the HDRTV format.

In this paper, we present a method to address the challenge of converting low-quality standard-dynamic-range television (LQ-SDRTV) to high-quality high-dynamic-range television (HQ-HDRTV) with a focus on improving the visual quality of the converted video. We propose a multi-frame content-aware mapping network, encompassing temporal-spatial alignment, feature modulation, and quality enhancement to significantly improve the performance of LQ-SDRTV to HQ-HDRTV conversion while simultaneously enhancing visual quality. Through the adoption of dynamic convolutions, hybrid prior extraction, and modulation modules, we demonstrate a robust and structured approach to handle the intricacies involved in SDRTV-to-HDRTV conversion, laying a solid foundation for further research and practical applications in this domain.

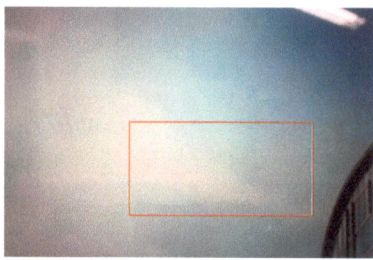

Figure 1. Amplified artifacts when applying the previous SDRTV-to-HDRTV method to low-quality SDR video. The notable artifacts are highlighted within the red rectangles. As shown in the left figure, HDR videos generated by the previous methods will contain amplified blocking artifacts. In the right figure, the banding artifacts are also amplified due to the lack of encoding compression.

The main contributions can be summarized as follows:

- We propose a multi-frame content-aware mapping network (MCMN) which takes into consideration the temporal continuity and spatial features of video frames in a structured manner to improve the performance from low-quality SDRTV to high-quality HDRTV.
- An innovative content-aware temporal spatial alignment module (CTAM) is introduced, employing dynamic deformable convolution to enhance the alignment accuracy of features across different frames and scales. Temporal spatial dynamic convolution (TSDC) adapts its convolution kernels based on the evolving temporal spatial patterns in the video, which is crucial for accurately capturing inter-frame relationships.
- The hybrid prior extraction module (HPEM) is designed to capture the multi-scale information in video content which is crucial for subsequent temporal spatial content-adaptive dynamic modulation.
- The temporal spatial transformation module (TSTM) employs a sequence of temporal spatial dynamic convolutions and mapping modules to perform content-adaptive dynamic modulation. Specifically, a cross-temporal mapping module (CTMM), a local spatial mapping module (LSMM), and a global spatial mapping module (GSMM) are introduced to refine both local and global details within images, leading to improved inverse tone mapping results and enhanced correction of encoding artifacts.

The rest of this paper is organized as follows. In Section 2, we introduce the related works to our proposed MCMN. Section 3 presents the motivation and detailed design of the MCMN for SDRTV-to-HDRTV artifact removal. The implementation and experimental results are demonstrated in Section 4. Finally, we conclude this paper in Section 5.

2. Related Work

2.1. SDRTV-to-HDRTV

Low dynamic range to high dynamic range (LDR-to-HDR) conversion methods aim to predict the physical brightness of a scene, allowing images to represent a broader spectrum of luminance. This is crucial for capturing scenes with significant light variation. Traditional techniques focus on estimating the light source density, which then aids in further expanding the dynamic range. Earlier methods [14–17] centered on estimating the light source density, using this as a foundation to broaden the dynamic range. Recent advancements have seen the application of deep learning, specifically deep convolutional neural networks, for this conversion. A notable method [18] introduced in 2020 directly converts LDR images to HDR without intermediate steps. Refs. [19,20] introduce the techniques that specifically target and recover overexposed areas in images. Another intriguing approach introduced in [21–24] offers a prediction mechanism. It predicts multi-exposure LDR image pairs using just a single LDR image. After this prediction, HDR images are synthesized based on the generated multi-exposure image pairs.

The SDRTV-to-HDRTV conversion approach has only emerged in the last two years. Ref. [8] proposes a GAN-based architecture that jointly achieves super-resolution and SDTV to HDRTV conversion. Ref. [9] proposes a hierarchical GAN architecture to accomplish super-resolution and SDTV to HDRTV conversion. Ref. [7] proposed a method using global feature modulation, local enhancement, and over-exposure compensation, which achieved the best performance. Ref. [10] proposed a global priors guided modulation network to extract color conformity priors and structural similarity priors that are beneficial for SDRTV-to-HDRTV and SR tasks, respectively. Similarly, the HDCFM framework [11] involves hierarchical dynamic context feature mapping to facilitate the learning of the mapping function from SDR frames to HDR frames. This is achieved through a hierarchical feature modulation module coupled with a dynamic context feature transformation module, providing a structured approach to understanding and transforming visual data from SDR to HDR and enhancing the visual output and potentially paving the way for better utilization of existing SDR video resources in the newer HDRTV format.

2.2. Artifact Removal

In terms of artifact removal task, numerous studies [25–35] are dedicated to improving the visual quality of compressed images and videos. ARCNN [25] was pioneering in its use of a CNN to alleviate image compression artifacts. Following this, Zhang et al. [26] introduced the DnCNN, which focused on various aspects of image restoration like denoising, deblocking, and super-resolution.

On the video quality enhancement front, Dai et al. [27] made the first stride by adapting a CNN for post-processing in HEVC intra coding. Building on this, Zhang et al. [28] aimed to replace HEVC's in-loop filter. On the other hand, Yang et al. [29] aimed to minimize distortion in HEVC by enhancing the quality of I and P/B frames without any encoder modifications. In a novel approach, He et al. [30] introduced a partition-masked CNN that leveraged coding unit size data to optimize network performance. Moreover, Ding et al. [31] proposed a squeeze-and-excitation filtering CNN which was designed as an optional in-loop filter to boost HEVC's efficiency. Xue et al. [32] devised a task-oriented flow network and employed motion compensation through a flow estimation module to enhance video quality. MFQE [33,34] recognized the significant fluctuations between compressed video frames and introduced a peak quality frame detector to improve videos' visual appearance. Lastly, to overcome the computational inefficiencies of optical flow, the spatio-temporal deformable fusion method [35] was presented as a solution for enhancing compressed videos.

3. Methodology

3.1. Overall

Our method can be primarily divided into three components: temporal-spatial alignment, tone mapping modulation, and quality enhancement. This approach aims to significantly improve the performance of converting low-quality SDRTV to high-quality HDRTV while simultaneously enhancing visual quality.

Given $2r + 1$ consecutive low-quality SDR frames $X_{[i-r:i+r]}$, we denote the center frame X_i as the target frame that needs to be mapped and the other frames as the reference frames. The input of the network is the target frame X_i and the $2r$ neighboring reference frames, and the output is the enhanced target frame Y_i^o. X is a stack of low-quality SDR frames which is defined as

$$X = [X_{i-r}, \cdots, X_{i-1}, X_i, X_{i+1}, \cdots, X_{i+r}], \tag{1}$$

where i denotes the frame index and r is the maximum range number of reference frames. The architecture of the MCMN is shown in Figure 2. In the following subsection, we will present a detailed analysis of the motivation behind and rationality of each module.

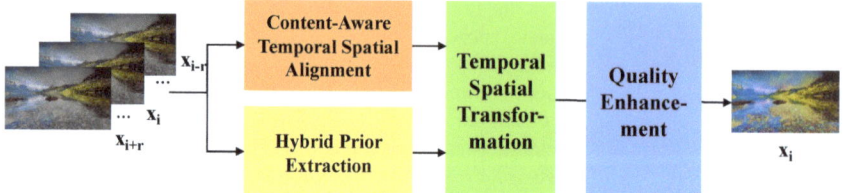

Figure 2. Architecture of our multi-frame content-aware mapping network (MCMN). Specifically, a series of low-quality SDR (LQ-SDR) frames are input to a content-aware temporal spatial alignment module to generate content-aware alignment features. Before proceeding with feature modulation, LQ-SDR frames are sent to the hybrid prior extraction module to yield triplet hybrid priors simultaneously. Next, with the help of a temporal spatial transformation module, the alignment features are tone-mapped into high-dynamic-range features with these hybrid priors. Finally, we employ a quality enhancement module to generate high-quality HDR results.

3.2. Content-Aware Temporal Spatial Alignment Module

When addressing the task of video conversion, we require a profound understanding of the temporal spatial correlation between adjacent video frames. To effectively harness this correlation, we introduce the content-aware temporal spatial alignment module (CTAM). The structure of the CTAM is shown in Figure 3. Considering the efficacy of prior deformable convolution [36] in video frame alignment tasks, we propose a dynamic deformable convolution (DDC) for the initial alignment in this task. We first employ the UNet [37] structure to extract feature offset with temporal spatial information, which can capture subtle changes in image features across different scales, thereby enhancing alignment accuracy. Specifically, the UNet-based offset extraction part consists of two down-sampling convolutional blocks, three convolutional blocks, and three up-sampling convolutional blocks. The number of neurons is 32 in all layers.

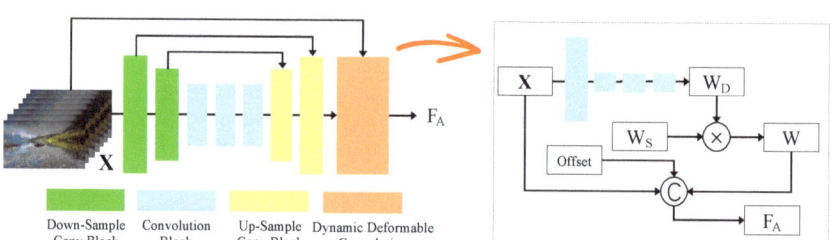

Figure 3. Structure of the content-aware temporal spatial alignment module (CTAM). By introducing learnable dynamic weights, dynamic deformable convolution is proposed to perform content-aware temporal spatial alignment.

To adaptively align these features from different temporal and spatial scales in a content-aware manner, we design a dynamic deformable convolution (DDC). Contrasting this with the standard deformable convolution [36], where only static weights are leveraged, DDC is engineered to determine content-aware characteristic features by introducing dynamic weights. This enables content-based temporal spatial alignment using the initial static weights. In DDC, the regular sampling grid $\mathcal{R} = \{(-1,-1),(-1,0),\cdots,(0,0),\cdots,(0,1),(1,1)\}$ is defined as the 3×3 convolution kernel with dilation 1. Here, the UNet-based offset extraction part is used to gather temporal spatial clues between X_{i-r} and X_{i+r} at diverse scales and generate the sampling parameters $\triangle P$ of the convolution kernel. It can be described as

$$\triangle P = UNet([X_{i-r},\cdots,X_{i-1},X_i,X_{i+1},\cdots,X_{i+r}]), \qquad (2)$$

where $\triangle P = \{\triangle p_k | k = 1, \cdots, |\mathcal{R}|\}$ is the collection of learnt predicted offsets $\triangle p_k$.

Many variants of DConv focus on finding better ways to enhance the offset. It should be noted that the weights of standard deformable convolution are static. Therefore, we introduce dynamic weights W_D which are learned from the original input directly. Here, we utilize cascaded global pooling, a 1×1 convolutional layer, and the sigmoid activation function to obtain the dynamic weights W_D. The content-adaptive weights are the combination of the static and dynamic weights in a dot multiplication. The aligned features F_{CTAM} from DDC at location p_0 can be computed:

$$F_{CTAM}(p_0) = \sum_{p_k \in \mathcal{R}} (W_k^S \otimes W_k^D) \cdot X(p_0 + p_k + \triangle p_k), \tag{3}$$

The convolution will be performed on the deformed sampling locations $p_k + \triangle p_k$, where W_k^S, W_k^D, p_k, and $\triangle p_k$ denote the static weights, dynamic weights, pre-specific offset, and learnt offset for k-th location in $\mathcal{R}$. $\otimes$ denotes dot multiplication.

After this, the content-adaptive temporal-spatial feature alignment features can be acquired. The success of this alignment process is crucial for subsequent tasks, as it aids in precisely capturing the temporal-spatial relationship between video frames, laying a solid foundation for feature modulation and recovery.

3.3. Hybrid Prior Extraction Module

In the field of video processing, features extracted from SDR video frames typically reside in the SDR feature space, while those extracted from HDR video frames are in the HDR feature space. As a result, the conversion from SDRTV to HDRTV can be modeled as a feature mapping. Correspondingly, previous works on SDRTV-to-HDRTV primarily focused on processing individual frames, obtaining low-dynamic-range features from SDR frames via a convolutional neural network, then mapping these features to high-dynamic-range features and finally restoring them to the HDR image space. Our approach is based on video, and fully exploits the spatio-temporal characteristics of videos. Hence, we introduce the spatio-temporal transformation module, aimed at capturing the spatio-temporal characteristic of videos, thereby achieving a superior conversion result.

Before proceeding with feature mapping, it is essential to extract mapping priors. To accommodate the unique requirements of our task, we design a hybrid prior extraction module (HPEM), specifically incorporating the cross-temporal prior extraction branch, local spatial prior extraction branch, and global spatial prior extraction branch. The detailed structure of the HPEM is shown in Figure 4. In the context of this module, we input the adjacent reference frames ($X_{i+t}, t \in [-r, \cdots, r], t \neq 0$) of the current target frame to the prior extraction module to obtain the cross-temporal prior F_{CTP}.

$$F_{CTP} = HPEM(X_{i+t}), \quad t \in [-r, \cdots, r], \quad t \neq 0, \tag{4}$$

where $HPEM$ denotes the hybrid prior extraction module. Specifically, five down-sample blocks are employed to generate down-sampled features F_D. Each block is composed of 1×1 conv, average pooling with a stride of 2, and LeakyReLU activation. The numbers of neurons in the five down-sampling blocks are 16, 32, 64, 128, and 128, respectively. For the cross-temporal prior F_{CTP}, adaptive pooling is used to yield the temporal weights.

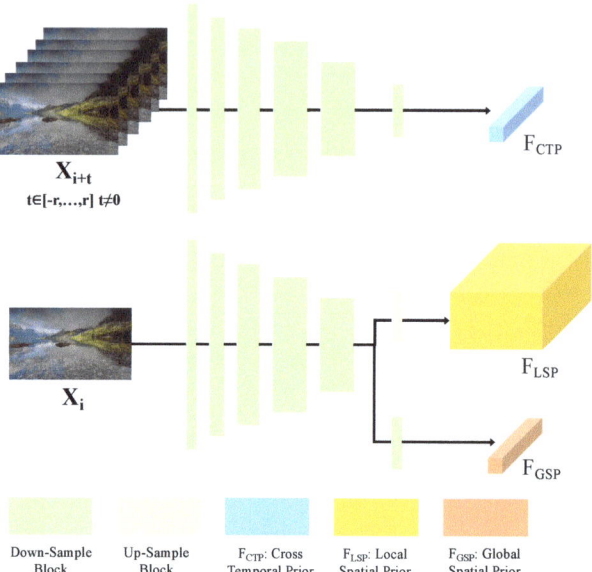

Figure 4. Structure of the hybrid prior extraction module (HPEM). In the cross-temporal prior branch, the adjacent reference frames ($X_{i+t}, t \in [-r, \cdots, r], t \neq 0$) are used to generate the cross-temporal prior using HEMP. In the local spatial and global spatial prior extraction branches, the local spatial prior and global spatial prior are generated using a single target center frame X_i simultaneously.

The current target frames are simultaneously individually processed by the HPEM to yield both the local spatial prior F_{LSP} and the global spatial prior F_{GSP}. The first part of this process involves distilling spatial information, and thus, the HPEM is employed to generate F_{LSP} and F_{GSP} simultaneously.

$$F_{LSP}, \ F_{GSP} = HPEM(X_i). \tag{5}$$

Specifically, for the local spatial prior (F_{LSP}), which targets pixel-level mapping, we employ a bilinear up-sampling operator to upscale the down-sampled features F_D to match the resolution of the input image. Conversely, for the global spatial prior (F_{GSP}), which is intended for frame-level mapping, we apply an average pooling operator to further down-sample the down-sampled features F_D. This process yields one learned global weight for each frame, facilitating effective global spatial mapping. This nuanced handling of spatial priors at different scales is critical for the fidelity of our spatial mapping operations.

The hybrid prior extraction module can adeptly capture multi-scale information of video content, preparing for the subsequent temporal spatial content-adaptive dynamic modulation.

3.4. Temporal Spatial Transformation Module

In the process of converting SDRTV to HDRTV, it is insufficient to process pixels from varied spatial locations identically due to the nuances they present. For example, a frame might contain both overexposed and underexposed areas, requiring tailored processing strategies for each exposure condition. To address this challenge, we develop a temporal spatial transformation module (TSTM) that is adaptive to spatio-temporal content. As shown in Figure 5, this method integrates a sequence of temporal spatial dynamic convolutions (TSDCs), a cross-temporal mapping module (CTMM), a local spatial mapping module (LSMM), and a global spatial mapping module (GSMM). Next, we will introduce each part in detail.

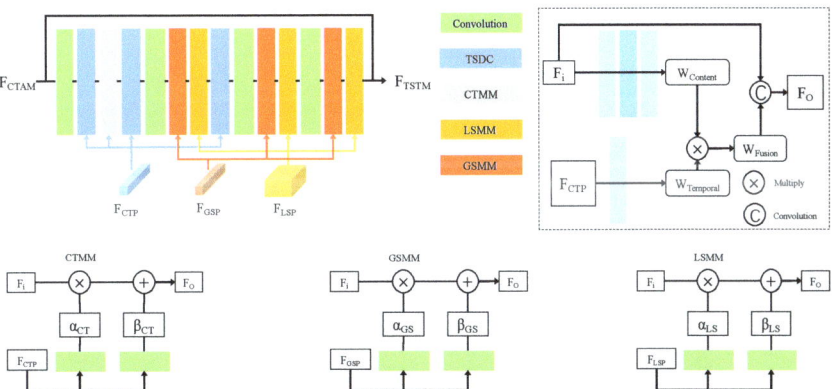

Figure 5. Structure of the temporal spatial transformation module (TSTM). It integrates a sequence of temporal spatial dynamic convolutions (TSDCs), a cross-temporal mapping module (CTMM), a local spatial mapping module (LSMM), and a global spatial mapping module (GSMM) to obtain better tone mapping results.

Temporal spatial dynamic convolution (TSDC) can adeptly modify its convolution kernels in response to the evolving spatio-temporal patterns in the video, aiding in capturing the inter-frame relationships more accurately. Specifically, we utilize the cross-temporal prior to generate the temporal weight $W_{Temporal}$ which contains a plethora of useful temporal-spatial information. The derived $W_{Temporal}$ is then fused with the original content weights $W_{Content}$ to obtain the fused weight W_{Fusion}. Finally, the input features are convoluted with the temporal spatial dynamically corrected weight to yield the feature output. This process can adaptively adjust the convolution kernel based on the input data to better capture the dynamic information within videos. The TSDC operator can be formulated by the following equation:

$$F_o(p_0) = \sum_{p_k \in \mathcal{R}} (W_{content} \otimes W_{temporal}) \cdot F_i(p_0 + p_k). \tag{6}$$

After TSDC, we propose the cross-temporal mapping module (CTMM) to perform tone mapping in the temporal dimension. Therefore, we revisit the spatial feature transform (SFT) [10,38] and global feature modulation (GFM) [7,39]. Inspired by the SFT and GFM, the proposed CTMM generates a pair of (α_{CT}, β_{CT}) by definition priors and performs modulations through scaling and shifting, respectively. The CTMM can be formulated by:

$$F_o(F_i|F_{CTP}) = \alpha_{CT} \otimes F_i \oplus \beta_{CT}, \tag{7}$$

where $\otimes$ refers to element-wise multiplication and $\oplus$ is element-wise addition. F_i and F_o are the input and output of the CTMM.

Moreover, spatial feature modulations, both local and global, focus on refining the details within images, leading to an improved inverse tone mapping result and a heightened ability to rectify encoding artifacts. The processes can be formulated by

$$\begin{aligned} F_o(F_i|F_{GSP}) &= \alpha_{GS} \otimes F_i \oplus \beta_{GS}, \\ F_o(F_i|F_{LSP}) &= \alpha_{LS} \otimes F_i \oplus \beta_{LS}, \end{aligned} \tag{8}$$

where the pairs of (α_{GS}, β_{GS}) and (α_{LS}, β_{LS}) by definition map and perform modulations through scaling and shifting in the global spatial mapping module and the local spatial mapping module.

3.5. Quality Enhancement Module

Encoding compression will lead to a deterioration in video quality, manifesting as blurring, block artifacts, or other visible compression distortions. To eliminate the artifacts by encoding compression, we introduce a quality enhancement module (QEM) at the final stage. Here, we take advantage of residual learning to generate the results. As illustrated in Figure 6, the module starts by applying a cascade of convolutional layers (here, we employ four layers) to extract the high-frequency information, denoted as the residual R_i^o. The final output, an enhanced high-quality HDR frame Y_i^o, is then obtained by residual learning. The process is formulated as follows:

$$R_i^o = QEM(F_{TSTM}),$$
$$Y_i^o = X_i + R_i^o. \qquad (9)$$

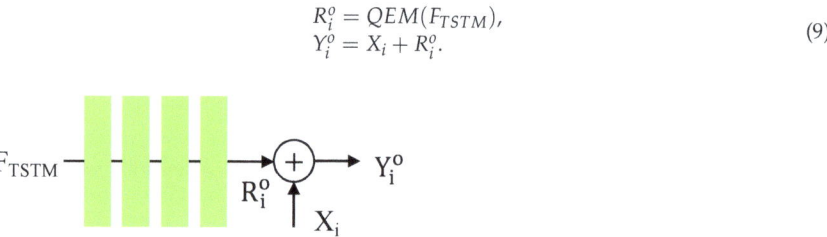

Figure 6. Structure of the quality enhancement module (QEM). We employ a cascade of convolutional layers with residual learning to yield the final enhanced high-quality HDR frame Y_i^o.

The primary idea of the QEM is to fully explore the complementary information within the fused feature maps and accordingly generate an enhanced high-quality HDR frame Y_i^o. Without bells and whistles, this straightforward QEM is capable of achieving satisfying enhancement results.

4. Results

In this section, we show the performance of the proposed MCMN. Section 4.1 introduces the training dataset and implementation details. In Section 4.2, we present the quantitative performance of the MCMN. Section 4.3 presents the qualitative performance. Moreover, the ablation study in Section 4.4 is demonstrated to prove the effectiveness of the designed architecture.

4.1. Experimental Settings
4.1.1. Dataset

We employ the well-used HDRTVNET dataset [7] as our benchmark. We use X265 [6] to encode SDR videos with different quantization parameters (QPs) (27, 32, 37, 42) to process the videos with different degrees of coding degradation. We compute the multi-scale structural similarity index (MS-SSIM) [40] of adjacent frames for scene segmentation. The MS-SSIM can evaluate the quality of video frames at multiple scales. This feature is crucial for capturing finer details and nuances that are especially relevant in HDR content.

4.1.2. Implementation Details

All experiments were conducted using PyTorch 1.6.0, Python 3.8, CUDA 10.1. The server was equipped with an Intel Core i9-13900K CPU and an NVIDIA GeForce RTX 4090 GPU. In our architecture, the range number r for the reference frame is 3. This configuration results in a total of seven input frames being considered for processing. Except for the previous special instructions, the number of neurons in all convolutional layers is 64. Moreover, the ReLU activation function is consistently used across these layers. During the training phase, we employ the SDR video with a quantization parameter (QP) of 37 as input data, and the output is a high-quality HDR video. The Adam optimizer, as mentioned in [41], is employed with an initial learning rate of 0.0005. After reaching 100,000 iterations, the learning rate is adjusted to halve every 60,000 iterations, totaling up to 660,000 iterations for the entire training process. We use L1 Loss between the high-quality high-dynamic standard image Y_i^{HQ-HDR} and the enhanced output Y_i^o of our MCMN to supervise network training. The loss function can be formulated as follows:

$$\mathcal{L} = \left\| Y_i^{HQ-HDR} - Y_i^o \right\|_1. \tag{10}$$

To verify the performance of different algorithms in a fair generalization, we take the last six stored weights to test the metrics. The model trained on coding degradation with a fixed QP = 37 was tested on four different QP coding test sets. As with [42], this multiple evaluation ensures that we can accurately and fairly evaluate the performance of different models.

4.2. Quantitative Results

The table presents a quantitative comparison of various image quality enhancement methods; specifically, it details their performance in terms of the peak signal-to-noise ratio (PSNR) and structural similarity index measure (SSIM) at different quantization parameter (QP) values. The PSNR is a widely used metric to measure the quality of reconstructed images, with higher values indicating a better quality. Similarly, the SSIM is another crucial measure that evaluates the perceptual quality of images by comparing structural similarities. We compare the proposed method with state-of-the-art SDRTV-to-HDRTV methods (FMNet [43], HDRUNET [44], HDCFM [11], HyCondITM [45], etc.). For a fair comparison, each method was retrained on the same training set. The last six checkpoints of each model were then tested on a common test set, and the average PSNR value was derived.

The quantitative results for each metric are shown in Table 1. The mean PSNR of our method is 33.208, while the mean PSNR of previous methods varies from 21.826 dB to 33.100 dB. It can be observed that our method consistently outperforms all comparison methods in terms of the mean PSNR for the test set, highlighting its robustness and superior performance. The difference in performance between the proposed method and the previous state-of-the-art (SOTA) method is notable. Specifically, comparing the proposed method with the preceding best performer, there is an improvement of 0.18 dB in the PSNR metric when the QP is set to 27. The proposed method showcases a superior performance in enhancing image quality across various quantization parameters.

Table 1. Quantitative comparison with previous methods. For each model on each QP, we selected the last six checkpoints for evaluation and calculated the mean PSNRs and SSIMs for comparison. The processing time (ms) was calculated for a resolution of 256 × 256. Best and second best results are **bold** and underlined, respectively.

Methods	Time	PSNR				Mean-PSNR	Mean-SSIM
		QP = 27	QP = 32	QP = 37	QP = 42		
CSRNET [39]	0.3	33.598	32.472	31.288	29.946	31.826	0.9518
STDF [35]	3.7	33.978	32.810	31.591	30.163	32.135	0.9455
AGCM [7]	0.4	34.260	33.123	31.878	30.395	32.414	0.9526
AILUT [46]	3.3	34.265	33.058	31.789	30.350	32.366	0.9498
DeepSRITM [8]	7.4	34.688	33.332	31.998	30.483	32.625	0.9515
FMNet [43]	1.4	34.462	33.474	32.146	30.584	32.666	0.9523
HDRUNET [44]	2.9	34.586	33.591	32.262	30.706	32.786	0.9514
HDCFM [11]	2.4	<u>34.897</u>	33.784	32.440	30.929	33.012	0.9538
HyCondITM [45]	8.3	34.860	<u>33.862</u>	<u>32.573</u>	**31.103**	<u>33.100</u>	<u>0.9554</u>
Ours	3.8	**35.072**	**34.026**	**32.651**	<u>31.083</u>	**33.208**	**0.9555**

Moreover, the superiority of our method extends to the SSIM as well, where it leads the pack among all the compared methods. This achievement in SSIMs, coupled with the high PSNR values, offers a more comprehensive assessment of our method's performance. By excelling in both objective metrics like the PSNR and subjective metrics like the SSIM, our method presents itself as a potentially pioneering solution in the realm of image quality enhancement, offering improvements from both objective and subjective perspectives.

4.3. Qualitative Results

Figure 7 presents the subjective comparison between our method and previous methods. In the first image, our approach can map the color of sunflower petals to the high-dynamic-range space with greater precision. Observing the second image, we can see that the artifacts in the lake section of the sample image are amplified by methods such as CSRNET [39], ACGM [7], HDCFM [11], and HyCondITM [45]. However, our technique effectively eliminates these artifacts caused by compression, bringing the image closer to the ground truth. In the third image, compared to other methods, our approach is able to restore the original image to a greater extent.

Figure 7. Qualitative results. Our method produces results with fewer artifacts and of a higher quality than previous methods (CSRNET [39], ACGM [7], AILUT [46], DeepSRITM [8], HDCFM [11], and HyCondITM [45]). Ground truth represents the reference standard against which the quality of our method's output is compared. The areas enclosed by the red rectangles are magnified for better observation.

4.4. Ablation Study

In this section, we will perform an ablation study to verify the effectiveness of the each part of our multi-frame content-aware mapping network (MCMN) design.

4.4.1. Ablation Study of the Content-Aware Temporal Spatial Alignment Module (CTAM)

To verify the effectiveness of our CTAM module, we conducted ablation studies. In this context, Exp. 1 denotes a baseline without any spatial temporal alignment, Exp. 2 denotes using deformable convolution for temporal alignment operations, and *MCMN* denotes a structure that utilizes dynamic deformable convolution for content-aware temporal spatial alignment. The results are demonstrated in Table 2. Without any temporal spatial alignment, as seen in the Exp. 1, the model's average PSNR stands at 32.390 dB. When the DCN is introduced for temporal alignment, there is a rise in the PSNR, reaching 32.689 dB. Most significantly, with the incorporation of the CTAM for content-aware temporal-spatial alignment in the model, the PSNR peaks at a value of 33.208 dB. This shows that our method effectively extracts spatial temporal information, aiding the model in tone mapping and artifacts.

Table 2. Ablation study of the content-aware temporal spatial alignment module (CTAM) in terms of the PSNR (dB). Best results are **bold**. ✓ indicates that the feature or component was enabled, while ✗ signifies that it was not enabled.

Exp.	Baseline	DCN	CTAM	Time	PSNR				Mean
					QP = 27	QP = 32	QP = 37	QP = 42	
1	✓	✗	✗	2.88	34.097	33.133	31.894	30.437	32.390
2	✓	✓	✗	2.95	34.595	33.409	32.137	30.617	32.689
MCMN	✓	✓	✓	3.83	**35.072**	**34.026**	**32.651**	**31.083**	**33.208**

4.4.2. Ablation Study of the Temporal Spatial Transformation Module (TSTM)

To validate the efficacy of individual components within the temporal spatial transformation module (TSTM), we undertook a comprehensive ablation study. This analysis spanned the global spatial mapping module (GSMM), the local spatial mapping module (LSMM), the cross-temporal mapping module (CTMM), and temporal spatial dynamic convolutions (TSDC), with a focus on adaptively reconstructing high-dynamic-range features.

Global spatial mapping module (GSMM) performance. When only the global spatial mapping module (GSMM) is employed (without the LSMM, the CTMM, and TSDC), the average PSNR is 32.390 dB. This serves as our baseline metric for performance.

Local spatial mapping module (LSMM) impact. Incorporating the LSMM alongside the GSMM (without the CTMM and TSDC) improves the average PSNR to 32.748 dB. This increment indicates the value of local spatial mapping in enhancing video quality.

Cross-temporal mapping module (CTMM) impact. Engaging the plain CTMM (only with the GSMM) further elevates the PSNR to 32.847 dB. This underscores the significance of temporal tone mapping. Notably, when introducing the LSMM simultaneously, the performance of the mean PSNR soared to 33.120 dB, marking a significant improvement of 0.273 dB. This finding reinforces the complementary nature of local and global and intra-frame and inter-frame information, collaboratively enhancing the method's performance.

Comprehensive module activation. The highest PSNR value of 33.208 dB was achieved when all modules (the GSMM, the LSMM, the CTMM, and TSDC) are utilized. This combined approach delivers the most optimal performance, with an impressive average PSNR increase of 0.818 dB compared to the baseline. Moreover, we conducted a qualitative comparison as part of our ablation study, shown in Figure 8. Compared with the ablation variants, our MCMN effectively eliminates artifacts and achieves color reproduction that closely matches the ground truth. This visual evidence, alongside the quantitative data, provides a more comprehensive evaluation of each part of our proposed method.

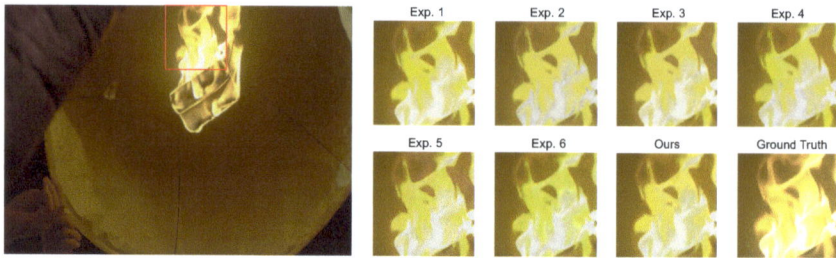

Figure 8. Qualitative results of the ablation study. Compared with the ablation variants (Exp. 1–6), our MCMN effectively eliminates artifacts and achieves color reproduction to closely match the ground truth. Ground truth represents the reference standard against which the quality of our method's output is compared. The areas enclosed by the red rectangle are magnified for better observation.

In conclusion, the temporal spatial transformation module (TSTM) plays a pivotal role in enhancing video quality during the SDRTV-to-HDRTV conversion process. Each module within the TSTM contributes uniquely to this enhancement, with their collective use delivering the most superior results.

4.5. Processing Time

We evaluated the processing time required for converting SDRTV using our method. All measurements were conducted on an NVIDIA GeForce RTX 4090Ti GPU to ensure a consistent testing environment. The processing time was calculated for a resolution of 256 × 256. As detailed in Table 1, our method achieves a processing time of 3.83 milliseconds (ms), corresponding to a real-time processing speed of approximately 261 frames per second (fps). Notably, our MCMN method is significantly more efficient, reducing the conversion time by over 50% compared to the previous SOTA algorithm, HyCondITM. This marked improvement in processing speed, coupled with enhancements in video quality, highlights the practical benefits of our approach.

Furthermore, we analyzed the processing times for various ablation study variants. Tables 2 and 3 illustrate that while adding each module increases the processing time, the corresponding performance enhancements are substantial. This demonstrates a favorable trade-off between processing time and performance improvements.

Table 3. Ablation study of the temporal spatial transformation module (TSTM) in terms of the PSNR (dB). Best results are **bold**. ✓ indicates that the feature or component was enabled, while ✗ signifies that it was not enabled.

Exp.	GSMM	LSMM	CTMM	TSDC	Time	PSNR QP = 27	QP = 32	QP = 37	QP = 42	Mean
3	✓	✗	✗	✗	3.09	34.097	33.133	31.894	30.437	32.390
4	✓	✓	✗	✗	3.49	34.628	33.498	32.195	30.673	32.748
5	✓	✗	✓	✗	3.43	34.865	33.567	32.252	30.704	32.847
6	✓	✓	✓	✗	3.56	34.977	33.923	32.569	31.012	33.120
MCMN	✓	✓	✓	✓	3.83	**35.072**	**34.026**	**32.651**	**31.083**	**33.208**

5. Conclusions

In conclusion, this paper delineates a comprehensive and innovative methodology for addressing the nuanced task of converting low-quality standard-dynamic-range television (LQ-SDRTV) to high-quality high-dynamic-range television (HQ-HDRTV). The focal point of this endeavor is not only to achieve a superior conversion quality but also to significantly enhance the visual quality of the resultant videos, addressing the prevalent issues associated with existing methods. The proposed multi-frame content-aware mapping network

(MCMN) epitomizes a holistic approach towards this objective, including temporal-spatial alignment, feature modulation, and quality enhancement within a coherent framework.

The introduction of the content-aware temporal spatial alignment module (CTAM) underscores a crucial step towards accurately aligning features across various scales and frames, facilitated by dynamic deformable convolution. Combining this with temporal-spatial dynamic convolution (TSDC) lays a robust foundation for capturing the intricate inter-frame relationships inherent in video sequences. The hybrid prior extraction module (HPEM) and the temporal spatial transformation module (TSTM) further accentuate the methodical approach employed in this work. These modules diligently capture multi-scale information and perform content-adaptive dynamic modulation, respectively, thereby addressing both local and global details within images. The resultant enhanced inverse tone mapping and the correction of encoding artifacts signify notable advancements in the SDRTV to HDRTV conversion domain.

The proposed methodology not only showcases the flexibility and scalability of the approach but also heralds a promising avenue for future research and practical applications. Each module within the network embodies a targeted strategy to tackle specific challenges associated with SDRTV to HDRTV conversion, making the methodology adaptable and conducive for further refinements.

From time to space, from local to global, our method makes full use of multi-frame information to perform inverse tone mapping of single-frame images, while it is also able to better repair coding artifacts. The results emanating from this work underscore a significant stride towards bridging the gap between SDRTV and HDRTV technologies, making a compelling case for the adoption and further exploration of the proposed methodology in real-world applications.

Author Contributions: Conceptualization, Z.W. and G.H.; methodology, Z.W.; software, Z.W.; validation, Z.W. and G.H.; formal analysis, Z.W.; investigation, G.H.; resources, G.H.; data curation, Z.W.; writing—original draft preparation, Z.W.; writing—review and editing, G.H.; visualization, Z.W.; supervision, G.H.; project administration, G.H.; funding acquisition, G.H. All authors have read and agreed to the published version of the manuscript.

Funding: This work is supported by project CEIEC-2022-ZM02-0247.

Institutional Review Board Statement: Not applicable.

Informed Consent Statement: Not applicable.

Data Availability Statement: Data are contained within the article.

Conflicts of Interest: The authors declare no conflicts of interest.

References

1. Mantiuk, R.; Daly, S.; Kerofsky, L. Display adaptive tone mapping. In Proceedings of the SIGGRAPH '08: ACM SIGGRAPH 2008 Papers, Los Angeles, CA, USA, 11–15 August 2008; pp. 1–10.
2. *SMPTE ST 2084*; High Dynamic Range Electro-Optical Transfer Function of Mastering Reference Displays. SMPTE: White Plains, NY, USA, 2014; p. 11.
3. Nagata, Y.; Ichikawa, K.; Yamashita, T.; Mitsuhashi, S.; Masuda, H. Content Production Technology on Hybrid Log-Gamma. In Proceedings of the SMPTE 2017 Annual Technical Conference and Exhibition, Los Angeles, CA, USA, 23–26 October 2017; pp. 1–12.
4. Rissanen, J.; Langdon, G.G. Arithmetic coding. *IBM J. Res. Dev.* **1979**, *23*, 149–162. [CrossRef]
5. Chen, Y.; Murherjee, D.; Han, J.; Grange, A.; Xu, Y.; Liu, Z.; Parker, S.; Chen, C.; Su, H.; Joshi, U.; et al. An overview of core coding tools in the AV1 video codec. In Proceedings of the 2018 Picture Coding Symposium (PCS), San Francisco, CA, USA, 24–27 June 2018; pp. 41–45.
6. Sullivan, G.J.; Ohm, J.R.; Han, W.J.; Wiegand, T. Overview of the high efficiency video coding (HEVC) standard. *IEEE Trans. Circuits Syst. Video Technol.* **2012**, *22*, 1649–1668. [CrossRef]
7. Chen, X.; Zhang, Z.; Ren, J.S.; Tian, L.; Qiao, Y.; Dong, C. A New Journey From SDRTV to HDRTV. In Proceedings of the IEEE/CVF International Conference on Computer Vision (ICCV), Montreal, QC, Canada, 10–17 October 2021; pp. 4500–4509.
8. Kim, S.Y.; Oh, J.; Kim, M. Deep SR-ITM: Joint Learning of Super-Resolution and Inverse Tone-Mapping for 4K UHD HDR Applications. In Proceedings of the International Conference on Computer Vision, Seoul, Republic of Korea, 27–28 October 2019.

9. Kim, S.Y.; Oh, J.; Kim, M. Jsi-gan: Gan-based joint super-resolution and inverse tone-mapping with pixel-wise task-specific filters for uhd hdr video. In Proceedings of the AAAI Conference on Artificial Intelligence, New York, NY, USA, 7–12 February 2020; Volume 34, pp. 11287–11295.
10. He, G.; Long, S.; Xu, L.; Wu, C.; Yu, W.; Zhou, J. Global priors guided modulation network for joint super-resolution and SDRTV-to-HDRTV. *Neurocomputing* **2023**, *554*, 126590. [CrossRef]
11. He, G.; Xu, K.; Xu, L.; Wu, C.; Sun, M.; Wen, X.; Tai, Y.W. SDRTV-to-HDRTV via Hierarchical Dynamic Context Feature Mapping. In Proceedings of the 30th ACM International Conference on Multimedia (MM '22), New York, NY, USA, 10–14 October 2022; pp. 2890–2898. [CrossRef]
12. Xu, N.; Chen, T.; Crenshaw, J.E.; Kunkel, T.; Lee, B. Methods and Systems for Inverse Tone Mapping. U.S. Patent US9607364B2, 28 March 2017.
13. Ballestad Andrey, A.; Ward, K.J. Method and Apparatus for Image Data Transformation. U.S. Patent US9224363B2, 29 December 2015.
14. Akyüz, A.O.; Fleming, R.; Riecke, B.E.; Reinhard, E.; Bülthoff, H.H. Do HDR displays support LDR content? A psychophysical evaluation. *ACM Trans. Graph. (TOG)* **2007**, *26*, 38-es. [CrossRef]
15. Banterle, F.; Ledda, P.; Debattista, K.; Chalmers, A. Expanding low dynamic range videos for high dynamic range applications. In Proceedings of the 24th Spring Conference on Computer Graphics, Budmerice Castle, Slovakia, 21–23 April 2008; pp. 33–41.
16. Banterle, F.; Debattista, K.; Artusi, A.; Pattanaik, S.; Myszkowski, K.; Ledda, P.; Chalmers, A. High dynamic range imaging and low dynamic range expansion for generating HDR content. In *Computer Graphics Forum*; Wiley Online Library: Oxford, UK, 2009; Volume 28, pp. 2343–2367.
17. Marnerides, D.; Bashford-Rogers, T.; Debattista, K. Deep HDR hallucination for inverse tone mapping. *Sensors* **2021**, *21*, 4032. [CrossRef]
18. Liu, Y.L.; Lai, W.S.; Chen, Y.S.; Kao, Y.L.; Yang, M.H.; Chuang, Y.Y.; Huang, J.B. Single-image HDR reconstruction by learning to reverse the camera pipeline. In Proceedings of the IEEE/CVF Conference on Computer Vision and Pattern Recognition, Seattle, WA, USA, 13–19 June 2020; pp. 1651–1660.
19. Eilertsen, G.; Kronander, J.; Denes, G.; Mantiuk, R.K.; Unger, J. HDR image reconstruction from a single exposure using deep CNNs. *ACM Trans. Graph. (TOG)* **2017**, *36*, 178. [CrossRef]
20. Santos, M.S.; Ren, T.I.; Kalantari, N.K. Single image HDR reconstruction using a CNN with masked features and perceptual loss. *ACM Trans. Graph. (TOG)* **2020**, *39*, 1–10. [CrossRef]
21. Debevec, P.E.; Malik, J. Recovering high dynamic range radiance maps from photographs. In Proceedings of the 24th Annual Conference on Computer Graphics and Interactive Techniques, Los Angeles, CA, USA, 3–8 August 1997; pp. 369–378.
22. Lee, S.; An, G.H.; Kang, S.J. Deep recursive hdri: Inverse tone mapping using generative adversarial networks. In Proceedings of the European Conference on Computer Vision (ECCV), Munich, Germany, 8–14 September 2018; pp. 596–611.
23. Niu, Y.; Wu, J.; Liu, W.; Guo, W.; Lau, R.W. HDR-GAN: HDR image reconstruction from multi-exposed LDR images with large motions. *IEEE Trans. Image Process.* **2021**, *30*, 3885–3896. [CrossRef] [PubMed]
24. Yan, Q.; Zhang, L.; Liu, Y.; Zhu, Y.; Sun, J.; Shi, Q.; Zhang, Y. Deep HDR imaging via a non-local network. *IEEE Trans. Image Process.* **2020**, *29*, 4308–4322. [CrossRef] [PubMed]
25. Dong, C.; Deng, Y.; Change Loy, C.; Tang, X. Compression artifacts reduction by a deep convolutional network. In Proceedings of the IEEE International Conference on Computer Vision, Santiago, Chile, 7–13 December 2015; pp. 576–584.
26. Zhang, K.; Zuo, W.; Chen, Y.; Meng, D.; Zhang, L. Beyond a gaussian denoiser: Residual learning of deep cnn for image denoising. *IEEE Trans. Image Process.* **2017**, *26*, 3142–3155. [CrossRef] [PubMed]
27. Dai, Y.; Liu, D.; Wu, F. A convolutional neural network approach for post-processing in HEVC intra coding. In Proceedings of the International Conference on Multimedia Modeling, Reykjavik, Iceland, 4–6 January 2017; Springer: Berlin/Heidelberg, Germany, 2017; pp. 28–39.
28. Zhang, Y.; Shen, T.; Ji, X.; Zhang, Y.; Xiong, R.; Dai, Q. Residual highway convolutional neural networks for in-loop filtering in HEVC. *IEEE Trans. Image Process.* **2018**, *27*, 3827–3841. [CrossRef] [PubMed]
29. Yang, R.; Xu, M.; Liu, T.; Wang, Z.; Guan, Z. Enhancing quality for HEVC compressed videos. *IEEE Trans. Circuits Syst. Video Technol.* **2018**, *29*, 2039–2054. [CrossRef]
30. He, X.; Hu, Q.; Zhang, X.; Zhang, C.; Lin, W.; Han, X. Enhancing HEVC compressed videos with a partition-masked convolutional neural network. In Proceedings of the 2018 25th IEEE International Conference on Image Processing (ICIP), Athens, Greece, 7–10 October 2018; pp. 216–220.
31. Ding, D.; Kong, L.; Chen, G.; Liu, Z.; Fang, Y. A Switchable Deep Learning Approach for In-loop Filtering in Video Coding. *IEEE Trans. Circuits Syst. Video Technol.* **2019**, *30*, 1871–1887. [CrossRef]
32. Xue, T.; Chen, B.; Wu, J.; Wei, D.; Freeman, W.T. Video enhancement with task-oriented flow. *Int. J. Comput. Vis.* **2019**, *127*, 1106–1125. [CrossRef]
33. Yang, R.; Xu, M.; Wang, Z.; Li, T. Multi-frame quality enhancement for compressed video. In Proceedings of the IEEE Conference on Computer Vision and Pattern Recognition, Salt Lake City, UT, USA, 18–22 June 2018; pp. 6664–6673.
34. Guan, Z.; Xing, Q.; Xu, M.; Yang, R.; Liu, T.; Wang, Z. MFQE 2.0: A new approach for multi-frame quality enhancement on compressed video. *IEEE Trans. Pattern Anal. Mach. Intell.* **2019**, *43*, 949–963. [CrossRef] [PubMed]

35. Deng, J.; Wang, L.; Pu, S.; Zhuo, C. Spatio-temporal deformable convolution for compressed video quality enhancement. In Proceedings of the AAAI Conference on Artificial Intelligence, New York, NY, USA, 7–12 February 2020; Volume 34, pp. 10696–10703.
36. Dai, J.; Qi, H.; Xiong, Y.; Li, Y.; Zhang, G.; Hu, H.; Wei, Y. Deformable convolutional networks. In Proceedings of the IEEE International Conference on Computer Vision, Venice, Italy, 22–29 October 2017; pp. 764–773.
37. Ronneberger, O.; Fischer, P.; Brox, T. U-net: Convolutional networks for biomedical image segmentation. In Proceedings of the Medical Image Computing and Computer-Assisted Intervention–MICCAI 2015: 18th International Conference, Munich, Germany, 5–9 October 2015; Proceedings, Part III 18; Springer: Berlin/Heidelberg, Germany, 2015; pp. 234–241.
38. Wang, X.; Yu, K.; Dong, C.; Loy, C.C. Recovering realistic texture in image super-resolution by deep spatial feature transform. In Proceedings of the IEEE conference on Computer Vision and Pattern Recognition, Salt Lake City, UT, USA, 18–23 June 2018; pp. 606–615.
39. He, J.; Liu, Y.; Qiao, Y.; Dong, C. Conditional sequential modulation for efficient global image retouching. In Proceedings of the European Conference on Computer Vision, Glasgow, UK, 23–28 August 2020; Springer: Berlin/Heidelberg, Germany, 2020; pp. 679–695.
40. Wang, Z.; Simoncelli, E.P.; Bovik, A.C. Multiscale structural similarity for image quality assessment. In Proceedings of the Thrity-Seventh Asilomar Conference on Signals, Systems & Computers, Pacific Grove, CA, USA, 9–12 November 2003; Volume 2, pp. 1398–1402.
41. Kingma, D.P.; Ba, J. Adam: A method for stochastic optimization. *arXiv* **2014**, arXiv:1412.6980.
42. Ho, M.M.; Zhou, J.; He, G. RR-DnCNN v2.0: Enhanced Restoration-Reconstruction Deep Neural Network for Down-Sampling-Based Video Coding. *IEEE Trans. Image Process.* **2021**, *30*, 1702–1715. [CrossRef] [PubMed]
43. Xu, G.; Hou, Q.; Zhang, L.; Cheng, M.M. FMNet: Frequency-Aware Modulation Network for SDR-to-HDR Translation. In Proceedings of the 30th ACM International Conference on Multimedia (MM '22), New York, NY, USA, 10–14 October 2022; pp. 6425–6435. [CrossRef]
44. Chen, X.; Liu, Y.; Zhang, Z.; Qiao, Y.; Dong, C. HDRUnet: Single image HDR reconstruction with denoising and dequantization. In Proceedings of the IEEE/CVF Conference on Computer Vision and Pattern Recognition, Nashville, TN, USA, 20–25 June 2021; pp. 354–363.
45. Shao, T.; Zhai, D.; Jiang, J.; Liu, X. Hybrid Conditional Deep Inverse Tone Mapping. In Proceedings of the 30th ACM International Conference on Multimedia (MM '22), New York, NY, USA, 10–14 October 2022; pp. 1016–1024. [CrossRef]
46. Yang, C.; Jin, M.; Jia, X.; Xu, Y.; Chen, Y. AdaInt: Learning Adaptive Intervals for 3D Lookup Tables on Real-time Image Enhancement. In Proceedings of the IEEE/CVF Conference on Computer Vision and Pattern Recognition, New Orleans, LA, USA, 18–24 June 2022; pp. 17522–17531.

Disclaimer/Publisher's Note: The statements, opinions and data contained in all publications are solely those of the individual author(s) and contributor(s) and not of MDPI and/or the editor(s). MDPI and/or the editor(s) disclaim responsibility for any injury to people or property resulting from any ideas, methods, instructions or products referred to in the content.

Article

Block Partitioning Information-Based CNN Post-Filtering for EVC Baseline Profile

Kiho Choi [1,2]

[1] Department of Electronics Engineering, Kyung Hee University, Yongin-si 17104, Gyeonggi-do, Republic of Korea; aikiho@khu.ac.kr

[2] Department of Electronics and Information Convergence Engineering, Kyung Hee University, Yongin-si 17104, Gyeonggi-do, Republic of Korea

Abstract: The need for efficient video coding technology is more important than ever in the current scenario where video applications are increasing worldwide, and Internet of Things (IoT) devices are becoming widespread. In this context, it is necessary to carefully review the recently completed MPEG-5 Essential Video Coding (EVC) standard because the EVC Baseline profile is customized to meet the specific requirements needed to process IoT video data in terms of low complexity. Nevertheless, the EVC Baseline profile has a notable disadvantage. Since it is a codec composed only of simple tools developed over 20 years, it tends to represent numerous coding artifacts. In particular, the presence of blocking artifacts at the block boundary is regarded as a critical issue that must be addressed. To address this, this paper proposes a post-filter using a block partitioning information-based Convolutional Neural Network (CNN). The proposed method in the experimental results objectively shows an approximately 0.57 dB for All-Intra (AI) and 0.37 dB for Low-Delay (LD) improvements in each configuration by the proposed method when compared to the pre-post-filter video, and the enhanced PSNR results in an overall bitrate reduction of 11.62% for AI and 10.91% for LD in the Luma and Chroma components, respectively. Due to the huge improvement in the PSNR, the proposed method significantly improved the visual quality subjectively, particularly in blocking artifacts at the coding block boundary.

Keywords: EVC; MPEG-5; video coding standard; post-filtering; CNN

Citation: Choi, K. Block Partitioning Information-Based CNN Post-Filtering for EVC Baseline Profile. *Sensors* **2024**, *24*, 1336. https://doi.org/10.3390/s24041336

Academic Editors: Ittetsu Taniguchi, Jinjia Zhou and Xin Jin

Received: 18 January 2024
Revised: 7 February 2024
Accepted: 18 February 2024
Published: 19 February 2024

Copyright: © 2024 by the author. Licensee MDPI, Basel, Switzerland. This article is an open access article distributed under the terms and conditions of the Creative Commons Attribution (CC BY) license (https://creativecommons.org/licenses/by/4.0/).

1. Introduction

The current growth in global video applications, driven by consumer desire for high-quality experiences, has expanded the relevance of devices dramatically [1]. This spike has resulted in a significant increase in frame rates per second to support natural motion, leading to an increase in video content capacity. According to Cisco statistics [2], video-related traffic accounts for around 80% of overall Internet traffic, highlighting the widespread relationship between data transmission and video content. This trend is not confined to 2D movies; it includes 3D videos, volumetric stereoscopic images, 360-degree videos, and VR/AR material, all of which require more data capacity [3].

At the same time, the increase in Internet of Things (IoT) systems has increased the need for effective video coding technology [4]. The significant growth of video data in IoT systems pursues dedicated coding and processing methods. As these systems focus on local data processing for intelligent sensor nodes, the importance of minimizing data volume while ensuring high-quality decoded images becomes important [5]. Thus, video coding technology has importance in nodes that extend beyond traditionally used areas, including small edge units of computing.

Traditionally, video compression technology has been developed through standards created by organizations such as the ISO/IEC Moving Picture Experiences Group (MPEG) and ITU-T Video Coding Experiences Group (VCEG). Standards such as MPEG-2/H.262 [6], Advanced Video Coding (AVC)/H.264 [7], and High Efficiency Video Coding (HEVC)/H.265 [8] have contributed significantly to the efficient compression and transmission of video data. Recently, new video coding standards such as Versatile Video Coding (VVC)/H.266 [9] and MPEG-5 Essential Video Coding (EVC) [10] have been introduced. While VVC/H.266 was developed jointly by the MPEG and VCEG, EVC is a product exclusively for the MPEG.

In this context of the growth of video data in IoT systems, it is necessary to carefully review the completed EVC standard. In particular, the EVC Baseline profile is customized to meet the specific needs of handling IoT video data. Since this profile aims to build a royalty-free codec using conventional coding techniques that are more than 20 years old, focusing only on performing key functions, avoiding the integration of complex tools, results in high-performance compression even at a low complexity [11]. Therefore, it is believed that the EVC Baseline profile proves to be a proper video codec for sensor node networks that require high-performance compression while operating at low power and complexity, and it is expected to play a pivotal role in addressing the growing need for high performance within the IoT ecosystem.

However, the EVC Baseline has a notable drawback. Being a codec comprised solely of simple tools developed over 20 years, it tends to exhibit numerous coding artifacts. Specifically, the presence of blocking artifacts at the block boundaries is considered a critical issue that needs resolution. To address this, a post-filter leveraging block partitioning information-based Convolutional Neural Network (CNN) is introduced in this paper. The proposed filter aims to rectify the challenges associated with the EVC Baseline profile, characterized by a high occurrence of coding artifacts. The proposed post-filter seeks to provide a high-efficiency compression performance and enhanced image quality, making it suitable for node sensor networks with low-complexity requirements. The main contributions of this study can be summarized as follows:

(1) A CNN-based post-filter for the EVC Baseline profile was developed, offering a promising video coding solution for IoT devices.
(2) An analysis of the major artifacts in the EVC Baseline profile was conducted, and a method indicating the area where these artifacts appear was exploited.
(3) The incorporation of a guide map based on blocking partitioning information was implemented to identify attention areas and enhance visual quality in the target image and video.
(4) Consideration was given to IoT applications with low complexity, allowing IoT devices to selectively add the post-filter based on the available extra computing power.
(5) A scenario-based CNN-based post-processing network was developed for real IoT applications, whether in image-based or real-time broadcasting/streaming services.

The remainder of this paper is organized as follows. Section 2 provides an overview of the EVC Baseline profile, related works of CNN-based filtering technologies, and standard activity. The proposed method is presented in Section 3. Section 4 provides an overall performance evaluation and analysis. Finally, Section 5 concludes this paper.

2. Related Work

To examine the relevant work of the proposed method, this section initially offers background information on EVC Baseline profiles. Following that, it explores CNN-based filtering for video coding, encompassing both in-loop filtering and out-loop filtering. Finally, it will outline the recent developments in standards for neural network-based video coding at the Joint Video Exploration Team (JVET), a collaboration between the MPEG and ITU-T.

2.1. Overview of EVC Baseline Profile

The block structure of the EVC Baseline profile is based on a partitioning method that supports quadtree division based on 64×64 blocks. The maximum block size of the coding unit is 64×64 and the minimum size is 4×4. For intra prediction, the process is performed based on the coding unit block, incorporating five supported prediction modes. The intra prediction supported by the Baseline profile would be impossible to accurately predict direction, but the major intra directionality can be predicted, thereby reducing the redundancy of directional information within a frame.

The residual value generated in the prediction process is converted into a frequency value through Discrete Cosine Transform (DCT), and the converted coefficient value is converted into a quantized coefficient value through a quantization process. The size of the transformation aligns with the size of the prediction block, and a process with a quantization parameter (QP) in the range of 0 to 51 is used for quantization. After the quantization process, the quantized coefficient values are scanned through a zigzag scan order and are then binarized through a basic run-level coding method, and the binarized values are streamed to the entropy coding engine as described in JPEG Annex D [12]. In the case of the filtering tool in the Baseline profile, an initial version of the deblocking filter in AVC/H.264 was applied to improve the objective and subjective image quality. The method is the same as that applied to H.263 Annex J [13].

For the EVC Baseline profile, it has been reported that it achieves approximately 30% bit savings in the objective evaluation and about 40% in the subjective evaluation compared to AVC/H.264, which is widely utilized on the Internet, while maintaining the same quality [14]. Moreover, in terms of complexity, it exhibits one-fourth of the algorithmic complexity when compared to AVC/H.264, making it a promising candidate as an optimal compression codec for next-generation sensor nodes [15].

2.2. CNN-Based Filtering Technologies for Video Coding

To improve coding artifacts during the encoding and decoding process, the latest video coding standard comes equipped with an in-loop filter designed. In the case of the VVC/H.266 standard, it incorporates three traditional in-loop filters: Deblocking Filter (DBF), Sample Adaptive Offset (SAO), and Adaptive Loop Filter (ALF). These filters are sequentially applied to the reconstructed frames. The DBF focuses on suppressing blocking artifacts at the block boundaries, while the SAO filter and the ALF aim to eliminate artifacts resulting from quantization. Despite the effectiveness of these filters, there is still considerable room for improvement in terms of visual quality.

Recently, developments have seen an active pursuit of research aimed at minimizing video coding artifacts using neural networks. That research focuses primarily on two aspects: (1) the design of a filter using a neural network for an in-loop filtering method applicable within the codec, similar to the DBF, SAO, and ALF, and (2) the investigation of a post-filter method that can be selectively applied outside the codec as needed.

Park et al. [16] introduced a CNN-based In-Loop Filter (IFCNN) capable of replacing the SAO in HEVC/H.265, and the proposed IFCNN showed a promising coding performance on Bjontegaard Delta bitrate (BD-BR) [17], with reductions of 2.6% and 2.8% for the Random-Access (RA) and Low-Delay (LD) configurations, respectively. Dai et al. proposed a Variable Filter Size Residual Learning Convolutional Neural Network (VRCNN) [18], designed to replace conventional filters in HEVC/H.265, such as the DBF and SAO, in HEVC/H.265. The proposed method in [18] utilized the variable block size of transform in HEVC/H.265; thus, residual learning led to faster convergence. According to [18], the VRCNN reduced the BD-BR by an average of 4.6% in the All-Intra (AI) configuration. Similar to the motivation of earlier methods, Kang et al. introduced a multi-scale CNN (MMS-net) [19] that could replace the DBF and SAO in HEVC/H.265 by utilizing skip connections with different scales from subnetworks to enhance the restoration process. The proposed MMS-net's performance on the BD-BR showed a reduction of 8.5% for the AI configuration. Wang et al. [20] proposed an attention-based dual-scale CNN (ADCNN), which utilized the encoding information, such as the QP and partitioning information, and the proposed ADCNN's performance on the BD-BR showed reductions of 6.5% and 2.8% for the AI and RA configurations, respectively. The residual highway CNN (RHCNN) in [21] utilized residual units with a progressive training scheme for the QP bands, and the proposed RHCNN's performance on the BD-BR showed reductions of 5.7%, 4.4%, and 5.7% for the AI, RA, and LD configurations, respectively. Similar to the approach of [21], Wang et al. [22] applied a neural network-based in-loop filter (CNNLF) in the conventional video coding framework in VVC/H.266 by conducting the modules of feature extraction and image quality enhancement. Compared with VTM-15.0, the proposed CNNLF improved the PSNR by 0.4 dB and 0.8 dB at 0.1 Mbps, respectively, and by 0.2 dB and 0.5 dB at 1 Mbps, respectively. Huang et al. [23] also added the CNN-based network to the conventional video coding framework specifically between the DBF and SAO in VVC/H.266. The proposed method based on a variable CNN utilized an attention module into a residual block to extract informative features, and the proposed method in [23] showed reductions of 3.6%, 3.6%, and 4.2% in performance on the BD-BR for the AI, RA, and LD configurations, respectively.

The purpose for post-filtering approaches is similar to that of in-loop filtering; however, it is used outside of the codec architecture. Thus, CNN-based post-filtering algorithms can be selectively applied to decoded images to improve visual quality. Dong et al. [24] introduced a CNN-based artifact removal method (AR-CNN) designed for JPEG compressed images, which was an extension of the super-resolution CNN (SRCNN) from previous studies. The results presented in [24] demonstrated a 1 dB improvement achieved by the proposed AR-CNN when compared to JPEG images. Li et al. [25] presented a method employing a twenty-layer CNN architecture with residual learning. An interesting aspect of the method proposed in [25] involved transmitting side information related to video content complexity and quality indicators from the encoder to the decoder at each frame. The performance of the method, as reported in [25], demonstrated a 1.6% BD-BR reduction compared with HEVC/H.265 on the six sequences given in the 2017 ICIP Grand Challenge. Zhang et al. [26] introduced a post-processing architecture based on a CNN for VVC/H.266 compressed video sequences. This architecture utilized 16 identical residual blocks and incorporated three types of skip connections, and it was reported that the proposed method in [26] showed a reduction of 3.9% in performance on the BD-BR for the RA configuration compared to VVC/H.266. The authors extended the [26] method, incorporating a generative adversarial network (GAN)-based training strategy to improve the visual quality of VVC/H.266-decoded images. The proposed method in [27] showed a notable enhancement in perceptual visual quality, achieving a reduction of 3.9% in performance on the BD-BR for the RA configuration compared to VVC/H.266. Bonnineau et al. [28] introduced a multitask learning-based approach that employed a QP map to generalize the model with various QPs by sharing parameters within a single network and task-specific modules. The method presented in [28] exhibited a significant improvement in perceptual visual quality, achieving a reduction of 2.8% in performance on the BD-BR for the RA configuration compared to VVC/H.266. Wang et al. [29] aimed to enhance the visual quality of decoded images by incorporating partitioning information with QP information, introducing a three-branch network. The method described in [29] demonstrated a notable improvement in perceptual visual quality, achieving a reduction of 6.5% in performance on the BD-BR for the AI configuration compared to VVC/H.266. Meng et al. [30] presented a network for enhancing visual quality, combining temporal motion and spatial information through a fusion subnet and an enhancement subnet. The approach outlined in [30] showed a significant improvement in perceptual visual quality, achieving a 0.29 dB enhancement compared to VVC/H.266-decoded images.

2.3. Neural Network-Based Video Coding

Meanwhile, various applications have recently explored the advancement in neural network (NN) technology. For instance, machine learning is leveraged in natural language processing and computer vision to overcome performance barriers. This trend is also making an impact on the development of video coding. The JVET is actively monitoring the adoption of NN technology and has initiated research into Neural Network-based Video Coding (NNVC) [31]. During the 130th MPEG meeting and 19th JVET meeting, two independent Ad Hoc Groups (AHGs) related to NNVC were formed, both focusing on the development of (1) an end-to-end (E2E) video coding framework and (2) the integration of

NN in a hybrid video coding framework. Subsequently, these two AHGs were consolidated under the JVET, with the merged group tasked with assessing the feasibility of NNVC for potential coding gains compared to traditional video coding standards based on signal processing technology. Currently, the development of in-loop filtering mainly using neural networks is being actively discussed in the JVET. It should be noted that in JVET activities, the main architecture of the network is based on a res-block CNN structure. Considering the fact that video coding generally uses a residual-based encoding/decoding approach that relies on accurate predictions about intra/interframes, the focus for improvement is mainly on preserving the details expressed through content distribution without changing the DC value. Thus, this approach, using the res-block basis CNN architecture, aligns well with the overall architecture of video coding, proving effective for in-loop filtering.

3. CNN-Based Post-Filtering with Block Partitioning Information

In the previous section, we reviewed the filtering technologies employed in conventional video coding standards and the recently emerged neural network-based filtering methods. While the future outlook for neural network-based filtering technologies appears promising, it is acknowledged that they still present challenges in terms of complexity. Given this context, one might argue that a post-filter, capable of adaptively enhancing image quality as needed, is more practical than an in-loop filter, which must be consistently applied to sensor nodes requiring fast processing with low complexity. Therefore, this paper proposes a CNN-based post-filter for EVC, aiming to enhance the image quality and compression rates while maintaining the constraints of low power and low complexity.

3.1. Analysis of Coding Artifacts

The EVC Baseline profile employs a quadtree-based coding structure, allowing the utilization of blocks up to 64 × 64, as illustrated in Figure 1. This method involves determining the optimal block size through processing from 64 × 64 to 4 × 4 in the encoder and transmitting this information to the decoder based on the quadtree. For example, during the decoding process, if the split flag is 0, the coding block for the process is 64 × 64. If the split flag is 1, four additional split flags are transmitted, indicating whether the coding block should be divided into units of 32 × 32. This process continues until the information is transmitted down to 4 × 4, a leaf node. The size of the coding block is determined according to the characteristics of the content, and specifically, the coding block is determined as a large block in homogeneous areas and a small block in delicate areas. Nevertheless, while the EVC block decision process ensures optimal rate–distortion (RD) performance, the absence of high-performance in-loop filtering in the EVC Baseline profile leads to the generation of significant artifacts around the block.

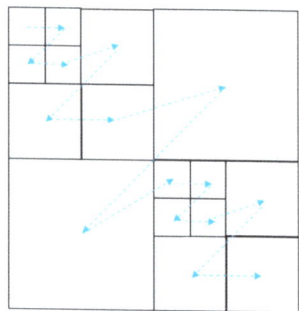
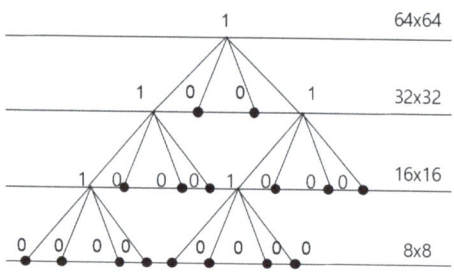

Figure 1. Quadtree-based coding structure in EVC Baseline profile.

Errors in the video coding process include ringing artifacts, blocking artifacts, and bending artifacts. Among these, the most noticeable artifact for video consumers is the blocking artifact, primarily occurring at the block boundaries in block-based video coding.

Specifically, in the EVC Baseline profile, the discontinuity at the block boundary is pronounced, leading to a significant degradation in the image quality of the decoded image. Figure 2 shows an example of the result of encoding by the EVC Baseline profile on QP = 37 to the RaceHorses sequence, clearly showing the prominent presence of blocking artifacts at the block boundary. The problem is that the Baseline profile contains an excessive number of such blocking artifacts. To address this concern, our research aims to improve the visual quality of the decoded images produced by the EVC Baseline profile. This improvement is accomplished by employing a block partitioning strategy within the context of CNN-based post-filtering.

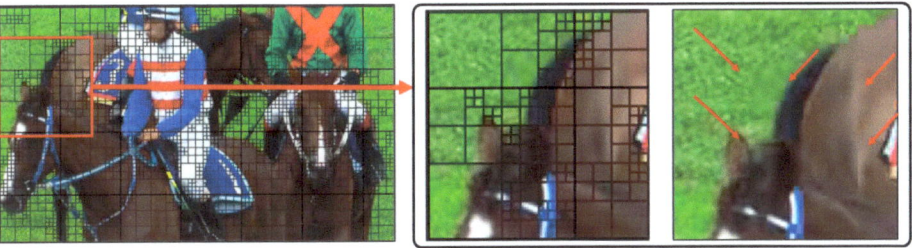

Figure 2. Example of coding artifacts detected in the area of the block boundary encoded with the EVC Baseline profile at the RaceHorses sequence with QP = 37.

3.2. Architecture and Network

Figure 3 depicts the overall pipeline for applying the proposed filtering in this paper. As depicted in the figure, in the case of this proposed post-filter, a filtering process is performed in the out-loop with the decoded image of the EVC Baseline profile. The CNN-based post-filter takes the decoded image and the block partitioning information extracted during the decoding process as the input and then improves the image quality by passing it through the trained CNN model.

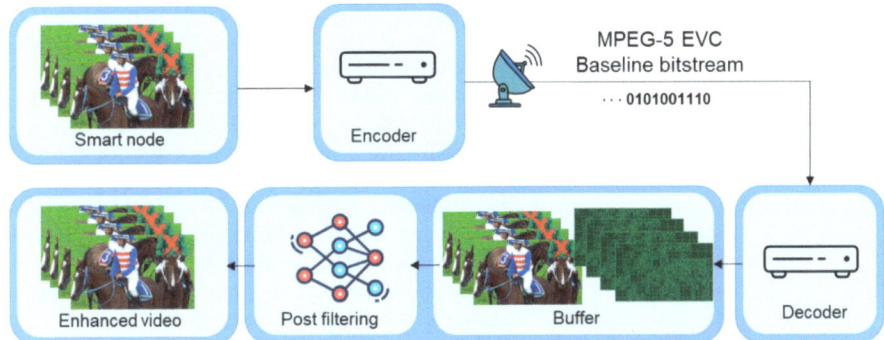

Figure 3. The overall pipeline for applying the proposed post-filtering in the use case.

The architecture in our proposed method is an extension of [32]. In the previous work, we utilized QP map information in the context of CNN-based post-filtering targeting the VVC/H.266 standard. In this paper, we extended this concept by integrating the block partitioning information into CNN-based post-filtering, specifically targeted for the EVC Baseline profile, which is suitable for video data transmission in sensor nodes. Figure 4 outlines the comprehensive network design employed in our proposed method.

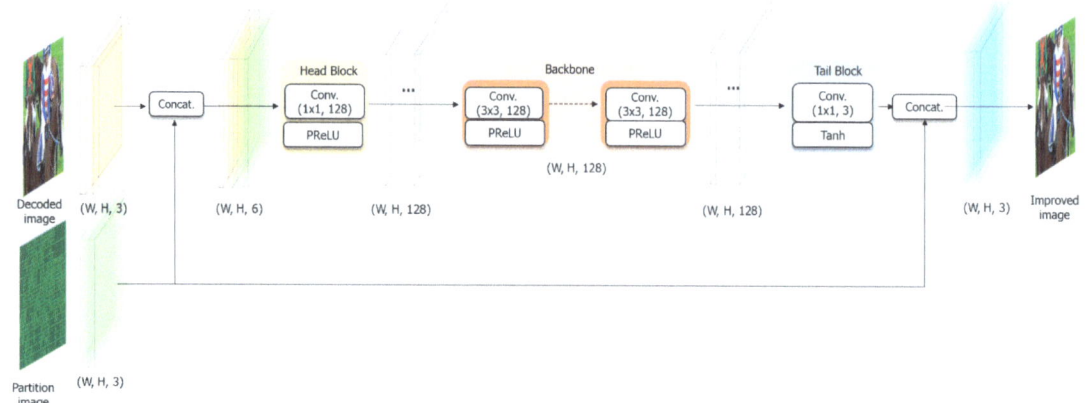

Figure 4. Proposed CNN-based post-filtering with block partitioning information.

In the initial processing block, the decoded image from the EVC Baseline profile is combined with a block partitioning map and framed. Both the decoded image from the EVC Baseline profile and the block partitioning map operate based on YUV channels. After concatenation, the number of input channels doubled. This package uses the decoded image from the EVC Baseline profile as the target for improvement, and the block partitioning map guides the areas with artifacts in the target decoded image.

The packaged video is then fed into the head block, comprising a 1×1 convolution filter with 128 output channels and a Parametric Rectified Linear Unit (PReLU). The primary function of the head block is to decrease the input dimensions for the subsequent backbone blocks. In this head block, we configured 128 channels to generate multiple feature representations, and the resulting 128 channels undergo the activation function (i.e., PReLU) to apply non-linearity to the output of the head block.

The features extracted from the head block are then directed to the backbone blocks, which encompass multiple blocks focused on extracting features. In the fundamental feature extraction block, a 3×3 convolution filter with 128 output channels and a PReLU is employed. The 3×3 convolution filter plays a crucial role in extracting features from the input and generating essential features for the subsequent layers. The 128 output channels from the 3×3 convolution filter undergo the PReLU activation function, and this process is repeated through the layers up to the final feature extraction block to ensure convergence in the deeper layers of the network.

In the proposed method described in this paper, we utilized 16 feature blocks based on empirical studies, but this number can be adjusted depending on the characteristics of the input decoded image. To maintain network simplicity, we designed the backbone block with a shape similar to the head block. While the head block primarily reduces the input dimensions, the backbone block focuses on capturing residual features for training.

The tail block, responsible for processing the output channels from the backbone blocks, integrates a 1×1 convolution filter with three output channels and employs the Tanh activation function, replacing the PReLU. To achieve precise quality improvement, we chose to update the residuals of the decoded image. As a result, the input decoded image from the head block is connected to the output of the tail block through a skip connection. The residual updates ensure that the primary values of the decoded image remain unaltered, while enabling adjustments to the corrupted areas introduced during the encoding and decoding processes, which represent the core objective of the proposed method.

3.3. Training

To create the training dataset, we utilized the BVI-DVC [33] dataset, comprised of 800 videos of varying resolutions ranging from 270 p to 2160 p, providing a diverse set of training data. Given that the BVI-DVC dataset is based on mp4 files, we converted these files to the YUV420 format with 10-bit files for the training dataset using FFmpeg [34]. To streamline the dataset creation, we extracted 10 frames from each video, resulting in a training dataset of 8000 frames. Ensuring uniform sizes for each Y, U, and V channel, we upsampled the U and V channels to match the size of the Y channel. Following the conversion, the original YUV format videos were processed through XEVE [35] and XEVD [36] to produce decoded images in the YUV format. Subsequently, instead of utilizing the entire image size for the training dataset, we cropped each image from the original and decoded YUV images to a size of 256 × 256. We then randomly selected the cropped images produced through horizontal and vertical flipping processes.

For the models in the proposed method, we generated five models corresponding to the QP values. Standard groups, such as the MPEG and JVET, use a common test condition (CTC) for experiments to evaluate suggested contributions, usually utilizing four or five QP values. A QP value in a codec plays an important role in this process. An increase in the QP results in higher distortion due to a coarser quantization step applied to transform coefficients with a larger QP. This leads to the loss of high-frequency information and a broader distribution range for the compensation value between the reconstructed and original pixels. Conversely, a low QP value yields better visual quality but requires a relatively high bitrate. Therefore, the QP number serves as a fundamental control parameter determining the visual quality and bitrate of the video.

In the proposed method, the utilization of models dependent on the QP is a critical aspect contributing to the generation of high-quality outputs. During the training process, we generated five bitstreams and reconstructed YUV files, depending on the QP value, in accordance with the experiments carried out by the JVET CTC [31]. Subsequently, the proposed model was trained using these five bitstreams and reconstructed YUV files independently. Additionally, we generated models based on different configurations as well. Acknowledging that error characteristics vary with the QP and configuration, we developed a strategy to customize each model to specific error characteristics. The separate models for each scenario in the proposed approach are to ensure that the model is tailored to the specific requirements of each scenario. For instance, the AI model is trained for image-centric applications, while the LD model is designed for real-time broadcasting and streaming purposes. More detailed information on the training process is available in Table 1.

Table 1. Details of the training environment.

Training dataset	BVI-DVC
Videos	800 videos with 10 frames
Framework	Pytorch 1.13.0
Epoch	50
Optimizer	Adam optimizer with a learning rate of 10^{-4}
Models	Five models at QP22, 27, 32, 37, and 42 for AI Five models at QP22, 27, 32, 37, and 42 for LD
Anchor encoder	XEVE with Baseline profile setting
Anchor decoder	XEVD with Baseline profile setting
Hardware	AMD EPYC 7513 32-Core CPUs, 384 GB RAM (AMD, Santa Clara, CA, USA), and an NVIDIA A6000 GPU (NVIDIA, Santa Clara, CA, USA).

4. Experimental Results and Discussion

To evaluate the effectiveness of the proposed method, the JVET CTC [37] sequences were chosen for evaluation but were not included in the training dataset. These 19 sequences were classified into classes A1, A2, B, C, and D based on their resolution with characteristics. The test QP values for all configurations were 22, 27, 32, 37, and 42, corresponding to the JVET CTC. Given the potential applications for the EVC Baseline profile with CNN-based post-filtering, which could be used for a low-complexity, low-power sense node for video data transmission, we evaluated the proposed method using AI and LD configurations. Table 2 contains detailed test sequences and conditions.

Table 2. Details of the testing environment.

Test dataset	Class A1(4K): Tango2, FoodMarket4, Campfire Class A2(4K): CatRobot, DaylightRoad2, ParkRunning3 Class B(2K): MarketPlace, RitualDance, Cactus, BasketballDrive, BQTerrace Class C(WVGA): BasketballDrill, BQMall, PartyScene, RaceHorses Class D(WQVGA): BasketballPass, BQSquare, BlowingBubbles, RaceHorses
Frames	Full frames
Framework	Pytorch
Models	Five models at QP22, 27, 32, 37, and 42 for AI Five models at QP22, 27, 32, 37, and 42 for LD
Anchor encoder	XEVE with Baseline profile setting
Anchor decoder	XEVD with Baseline profile setting
Hardware	AMD EPYC 7513 32-Core CPUs, 384 GB RAM, and an NVIDIA A6000 GPU.

4.1. Objective Testing Result

For the objective evaluation, the increase in the PSNR was measured at the same bitrate of each sequence over all the QP values (i.e., BD-PSNR) [37], and the BD-BR was also measured to check on the bitrate reduction at the same visual quality that is usually used in standard experiments. Tables 3 and 4 present the experimental results of the proposed method for the AI and LD configurations, comparing the decoded image of the EVC Baseline profile to the enhanced image filtered by the CNN-based post-filter. The numbers in the table represent the average bitrate and PSNR of the five QPs of each sequence in the reference and proposed method.

Those metrics, the BD-PSNR and BD-BR, compare the improvement in the PSNR and coding efficiency of different video codecs or encoding settings while taking into consideration both the bitrate and video quality. The fundamental concept involved fitting a cubic polynomial curve through five data points and subsequently deriving an expression for the integral of the curve. The BD-PSNR allows for an objective assessment of PSNR improvement by calculating the difference in the PSNR to achieve a comparable bitrate between two codecs. In the BD-PSNR, a higher number indicates an improvement in the PSNR over the anchor. Similarly, by measuring the difference in the bitrate needed to attain an equivalent quality level between two distinct codecs, the BD-BR metric facilitates an objective assessment of the compression efficiency. The lower BD-BR value signifies a higher coding efficiency than the anchor at the same visual quality.

Table 3 shows the results of the proposed method compared to the reference in the AI configuration. As shown in Table 3, the proposed method increases the PSNR at the same bitrate by approximately 0.57 dB, 0.75 dB, and 0.95 dB for the Luma and Chroma components in the AI configuration when compared to the post-filter pre-processed video. The increased PSNR results in overall BD-BR reductions of 11.62%, 24.5%, and 28.79% for the Luma and Chroma components, respectively, in the AI.

Table 4 shows the results of the proposed method compared to the reference in the LD configuration. Similar results can be observed in the LD configuration. As shown in the table, the proposed method increases the PSNR at the same bitrate by approximately 0.37 dB, 0.82 dB, and 0.95 dB for the Luma and Chroma components in the LD configuration when compared to the post-filter pre-process video. The improved PSNR results in overall BD-BR reductions of 10.91%, 31.22%, and 32.30% for the Luma and Chroma components, respectively, in the LD.

Table 3. Objective testing result of AI configuration.

Class and Sequence		Bitrate (kpbs)	Reference (dB)			Proposed Method (dB)			BD-PSNR (ΔdB)			BD-BR (Δ%)		
			Y-PSNR	U-PSNR	V-PSNR	Y-PSNR	U-PSNR	V-PSNR	ΔY-PSNR	ΔU-PSNR	ΔV-PSNR	ΔY-BDBR	ΔU-BDBR	ΔU-BDBR
A1	Tango2	62,688	38.91	46.67	44.71	39.21	47.74	45.81	0.30	1.07	1.10	−11.42	−41.06	−38.15
	FoodMarket4	121,128	39.01	43.32	44.52	39.53	44.18	45.60	0.52	0.86	1.08	−12.49	−24.31	−29.93
	Campfire	76,616	37.58	39.26	40.26	37.83	40.51	41.07	0.25	1.25	0.81	−6.53	−33.13	−33.41
	CatRobot	122,884	37.73	40.35	40.85	38.27	41.08	41.89	0.53	0.73	1.04	−14.76	−36.53	−36.67
A2	DaylightRoad2	145,191	36.37	43.31	41.39	36.71	44.06	41.74	0.34	0.75	0.35	−11.83	−40.85	−21.43
	ParkRunning3	227,250	38.12	35.40	36.45	38.61	35.61	36.66	0.49	0.21	0.21	−8.34	−5.66	−7.67
B	MarketPlace	42,551	37.16	41.66	42.46	37.54	42.43	43.15	0.38	0.78	0.69	−9.35	−29.12	−28.89
	RitualDance	28,415	39.31	43.95	44.30	40.10	45.05	45.76	0.79	1.10	1.46	−15.27	−33.28	−39.10
	Cactus	47,502	35.36	38.73	40.63	35.84	39.11	41.41	0.48	0.38	0.78	−12.07	−17.94	−28.80
	BasketballDrive	31,843	36.65	42.18	42.70	37.09	42.29	43.31	0.43	0.11	0.61	−11.05	−5.96	−22.38
	BQTerrace	80,937	34.93	40.21	42.31	35.40	40.28	42.45	0.48	0.07	0.13	−7.96	−5.23	−9.76
C	BasketballDrill	11,741	35.27	39.88	39.93	36.25	40.69	41.63	0.98	0.81	1.70	−18.24	−25.33	−40.74
	BQMall	12,610	35.60	40.53	41.45	36.38	41.36	42.60	0.78	0.83	1.15	−14.26	−26.02	−32.56
	PartyScene	22,222	32.96	38.21	38.81	33.45	38.75	39.45	0.49	0.54	0.64	−8.19	−14.89	−16.41
	RaceHorses	7724	35.33	38.58	39.98	35.90	39.61	41.30	0.57	1.03	1.32	−11.12	−27.99	−38.47
D	BasketballPass	2895	35.85	40.78	40.28	36.67	41.92	41.68	0.81	1.14	1.39	−13.41	−28.42	−31.52
	BQSquare	7108	32.98	39.71	40.52	33.75	40.15	41.38	0.77	0.44	0.86	−10.45	−12.45	−23.37
	BlowingBubbles	5886	32.85	37.96	38.37	33.40	38.53	39.16	0.55	0.57	0.79	−9.57	−16.97	−21.52
	RaceHorses	2352	34.74	37.99	39.01	35.63	39.64	40.93	0.89	1.65	1.93	−14.38	−40.42	−46.22
	Average								0.57	0.75	0.95	−11.62	−24.50	−28.79

Table 4. Objective testing result of LD configuration.

Class and Sequence		Bitrate (kpbs)	Reference (dB)			Proposed Method (dB)			BD-PSNR (ΔdB)			BD-BR (Δ%)	
			Y-PSNR	U-PSNR	V-PSNR	Y-PSNR	U-PSNR	V-PSNR	ΔY-PSNR	ΔU-PSNR	ΔV-PSNR	ΔY-BDBR	ΔU-BDBR	ΔU-BDBR
A1	Tango2	36.94	45.84	43.38	37.17	46.69	44.26	0.23	0.23	0.85	0.88	−8.57	−59.49	−49.19
	FoodMarket4	35.95	41.03	41.92	36.19	42.28	43.40	0.25	0.25	1.26	1.48	−7.01	−63.93	−68.60
	Campfire	35.49	37.26	39.01	35.74	38.11	39.68	0.25	0.25	0.85	0.67	−7.28	−27.81	−34.49
	CatRobot	35.28	39.50	39.56	35.61	40.18	40.48	0.33	0.33	0.69	0.93	−10.55	−56.66	−49.27
A2	DaylightRoad2	33.97	41.81	39.97	34.16	42.75	40.67	0.19	0.19	0.94	0.70	−8.88	−71.13	−58.59
	ParkRunning3	34.05	33.09	34.51	34.30	33.33	34.82	0.26	0.26	0.24	0.31	−5.86	−12.42	−17.60
	MarketPlace	33.98	40.00	40.89	34.18	40.91	41.63	0.20	0.20	0.91	0.74	−6.59	−63.16	−56.89
	RitualDance	35.71	42.31	42.43	36.14	43.36	43.72	0.43	0.43	1.05	1.29	−9.36	−51.85	−54.47
B	Cactus	32.71	37.87	39.64	33.05	38.38	40.37	0.34	0.34	0.50	0.73	−11.87	−48.07	−44.86
	BasketballDrive	33.77	40.74	40.77	34.13	41.39	41.79	0.36	0.36	0.65	1.02	−11.47	−45.44	−48.28
	BQTerrace	31.19	37.83	39.75	31.52	38.72	40.86	0.32	0.32	0.89	1.11	−13.79	−65.73	−70.76
	BasketballDrill	31.83	37.89	37.72	32.43	39.11	39.20	0.60	0.60	1.22	1.48	−14.77	−48.76	−49.42
C	BQMall	31.73	38.74	39.58	32.23	39.87	40.91	0.50	0.50	1.13	1.32	−12.67	−57.01	−58.53
	PartyScene	28.39	36.39	37.13	28.74	36.97	37.67	0.35	0.35	0.58	0.55	−10.83	−29.20	−26.30
	RaceHorses	31.52	36.94	38.55	31.91	37.66	39.53	0.39	0.39	0.72	0.98	−10.72	−41.02	−54.39
	BasketballPass	31.88	39.26	38.18	32.44	39.91	39.10	0.57	0.57	0.65	0.92	−12.35	−27.59	−32.37
D	BQSquare	27.79	38.10	38.60	28.37	38.94	39.74	0.58	0.58	0.84	1.14	−18.51	−60.08	−68.07
	BlowingBubbles	28.37	35.93	36.49	28.68	36.54	37.01	0.31	0.31	0.60	0.52	−9.49	−31.72	−26.01
	RaceHorses	30.77	36.34	37.35	31.31	37.36	38.65	0.54	0.54	1.03	1.30	−12.68	−44.79	−52.78
	Average								0.37	0.82	0.95	−10.70	−47.68	−48.47

The objective test results show that the proposed method significantly improved the visual quality of the decoded images of the EVC Baseline profile, regardless of the QP or configuration. Notably, the proposed method outperforms the AI configuration in terms of a low resolution and high QP values. The main reason for the significant improvements can be attributed to areas where blocking artifacts noticeably appear. Blocks in the AI configuration typically determine the number of coding blocks with a small coding block size; increasing the number of coding blocks results in more blocking artifacts in the block boundary area. This phenomenon can also be applied to the output of low-resolution sequences. Because of the high amount of quantized values for en/decoding, the artifact can be widely visible in coding blocks regardless of the coding block size or number. As a result, the improvements at high QP sequences would be due to an improvement in visual quality across the entire decoded image.

4.2. Subjective Testing Result

To assess the improvement in visual quality achieved by our proposed method, individual visual quality evaluations were conducted. Figures 5 and 6 present a comparative analysis of the visual quality for the AI and LD configurations between the decoded image of the EVC Baseline profile and the proposed results. The visual quality assessment was performed at a middle QP value as QP = 32. Figure 5 illustrates the comparison results for PartyScene with the AI configuration. The filtered image by post-filtering reveals a superior visual quality compared to the video before post-filtering, aligning with the 0.49 dB improvement observed in the objective evaluation. Notably, the figure of the proposed method in Figure 5 shows a further reduction in artifacts, especially in the face of the child and around the area of the boxes. Similar results are evident in the LD configuration. Figure 6 shows the comparison results for BQTerrace with the LD configuration. In this figure, the filtered image by the proposed method reveals more textural detail than the reference, especially notable in the parasol, where texture lines are clearly observed in the proposed filtered image.

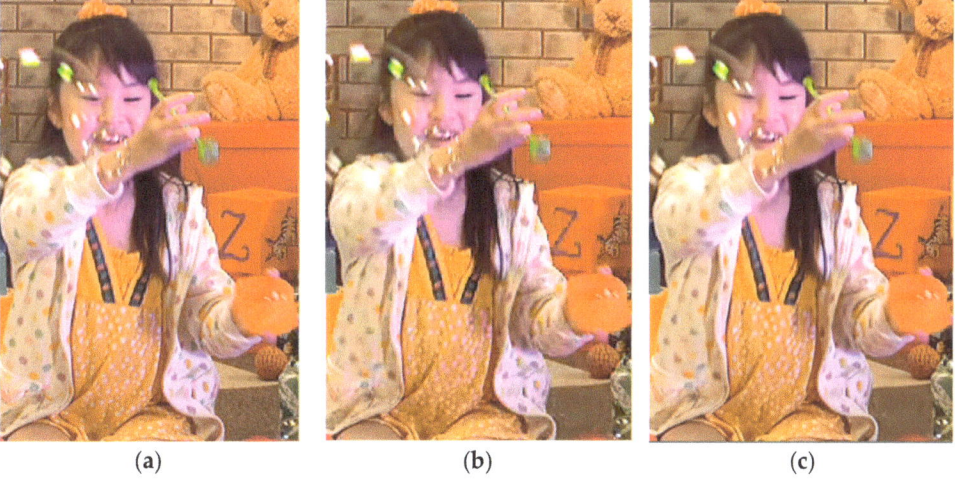

Figure 5. Visual quality comparison with AI configuration at PartyScene with #0 frame: (**a**) original image, (**b**) decoded image, (**c**) proposed method.

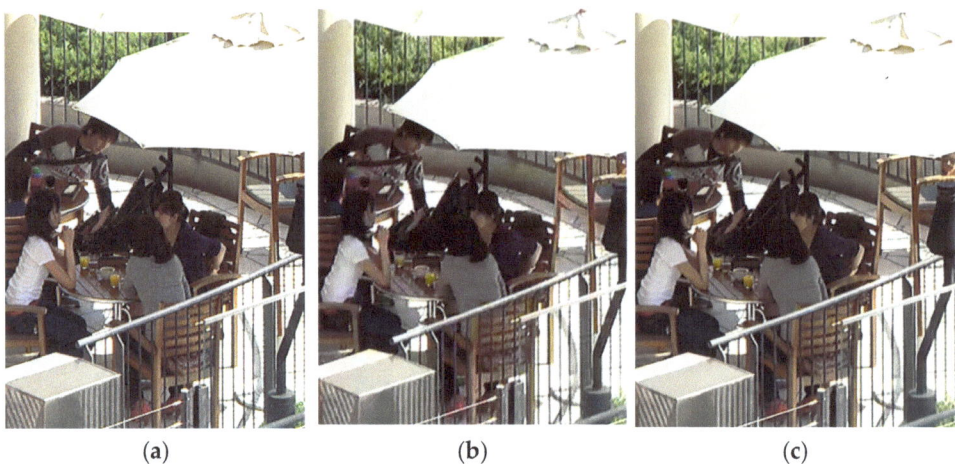

Figure 6. Visual quality comparison with LD configuration at BQTerrace with #11 frame: (**a**) original image, (**b**) decoded image, (**c**) proposed method.

4.3. Discussion

The experimental results of the proposed method showed both objective and subjective improvements in performance. In the objective results, the PSNR showed a significant enhancement in both the AI and LD configurations. This PSNR improvement corresponds to a reduced bitrate at the same image quality. Considering the historical fact of video coding standards improving performance by about 50% every decade, achieving a 10% enhancement with just one tool is quite impressive objectively. The subjective experimental results also reveal a remarkable improvement in image quality. The proposed method effectively addresses blocking artifacts, a specific target of enhancement, noticeably eliminating them. Another noteworthy aspect in the subjective experiments is the preservation of details in the original video that typically disappear due to coding artifacts. The proposed method successfully restores these lost details, bringing the visual quality closer to the original video. The significant improvement in visual quality is attributed to the CNN-based post-filter guided by the partitioning map, which identifies areas affected by blocking artifacts and guides the CNN model to enhance these areas in line with the original video.

Nevertheless, it should be noted that implementing the proposed CNN-based method involves considerable complexity, particularly when applied directly to devices within the IoT. The decision to employ a post-filter for enhancing image quality, accounting for diverse computing performances, aligns with the practical challenge of deploying CNN-based filters across the spectrum of IoT devices. In this context, it is believed to be more pragmatic to adaptively apply these filters as add-ons using post-filtering instead of in-loop filtering when external resources are available, as illustrated in the proposed configuration.

4.4. Future Work

This paper introduced a CNN-based post-filtering method designed for the EVC Baseline profile to address the requirements of IoT devices. While the proposed method is specifically applied to the EVC Baseline profile, its foundational architecture, shared with other video coding standards, suggests its potential applicability in the EVC Main profile or diverse video codecs. The exploration of extending the method outlined in this paper to other video codecs is considered an interesting topic for future research.

Additionally, it is noted that CNN remains an important role in enhancing the coding performance in this paper. With NNVC in the JVET successfully employing CNN-based deep-learning technology, a similar investigation into CNN-based filtering technology for the EVC Baseline profile has been conducted in the proposed method. Given the novelty of the MPEG-5 EVC Baseline profile and the limited research on leveraging CNN as a post-filter for this codec, exploring this aspect in conjunction with the latest machine-learning techniques is also considered an interesting topic for future research.

5. Conclusions

In this paper, a post-filter utilizing a CNN with block partition information for the EVC Baseline profile was proposed. As the demand for efficient video coding technology intensifies, driven by the surge in video data from IoT devices, the EVC Baseline can be considered as a promising solution designed to address the specific requirements of processing IoT video data with low complexity. Nevertheless, enhancements are required to address coding artifacts within the EVC Baseline profile. To tackle this issue, a post-filter utilizing a CNN based on block partition information was introduced in this paper. Through experimental results, both objective and subjective assessments showed significant improvements in both the AI and LD configurations when compared to the pre-post-filter video. The advancements achieved by the proposed method notably enhanced the visual quality, especially in blocking artifacts at the boundaries. Thus, this proposed method is expected to benefit networks of high-performance, low-complexity sensor nodes in IoT ecosystems using the EVC Baseline profile.

Funding: This work was supported by a grant from Kyung Hee University in 2023 (KHU-20230874) and was supported by the Institute of Information & communications Technology Planning & Evaluation (IITP) grant funded by the Korea government (MSIT) (RS-2023-00220204, Development of Standard Technologies on next-generation media compression for transmission and storage of Metaverse contents).

Institutional Review Board Statement: Not applicable.

Informed Consent Statement: Not applicable.

Data Availability Statement: Data are contained within the article.

Conflicts of Interest: The author declares no conflicts of interest.

References

1. Zhang, C. The why, what, and how of immersive experience. *IEEE Access* **2020**, *8*, 90878–90888. [CrossRef]
2. Cisco Visual Networking Index: Forecast and Methodology, 2016–2021. Available online: http://www.cisco.com/c/en/us/solutions/collateral/service-provider/visual-networking-index-vni/complete-white-paper-c11-481360.html (accessed on 10 October 2023).
3. Augmented Reality and Virtual Reality Market with COVID-19 Impact Analysis by Offering (Hardware & Software), Device Type (HMD, HUD, Gesture Tracking), Application (Enterprise, Consumer, Commercial, Healthcare), and Geography—Global Forecast to 2025. Available online: https://www.marketsandmarkets.com/Market-Reports/augmented-reality-virtual-reality-market-1185.html (accessed on 10 October 2023).
4. Mukhopadhyay, S.C.; Tyagi, S.K.S.; Suryadevara, N.K.; Piuri, V.; Scotti, F.; Zeadally, S. Artificial intelligence-based sensors for next generation IoT applications: A review. *IEEE Sens. J.* **2021**, *21*, 24920–24932. [CrossRef]
5. Chen, C.W. Internet of video things: Next-generation IoT with visual sensors. *IEEE Internet Things J.* **2020**, *7*, 6676–6685. [CrossRef]
6. *ISO/IEC 13818-2 MPEG-2 and ITU-T Recommendation H.262*; Generic Coding of Moving Pictures and Associated Audio Information—Part 2: Video. International Telecommunication Union: Geneva, Switzerland, 1994.
7. *ITU-T Recommendation H.264 and ISO/IEC 14496-10*; Advanced Video Coding (AVC). International Telecommunication Union: Geneva, Switzerland, 2003.
8. *ITU-T Recommendation H.265 and ISO/IEC 23008-2*; High Efficient Video Coding (HEVC). International Telecommunication Union: Geneva, Switzerland, 2013.
9. *ITU-T Recommendation H.266 and ISO/IEC 23090-3*; Versatile Video Coding (VVC). International Telecommunication Union: Geneva, Switzerland, 2020.
10. *ISO/IEC 23094-1:2020*; Information Technology—General Video Coding—Part 1: Essential Video Coding. International Organization for Standardization: Geneva, Switzerland, 2020.
11. Choi, K.P.; Choi, K.; Park, M.W.; Park, M.; Piao, Y.; Choi, M.; Yang, H.; Park, Y. Overview of baseline profile in MPEG-5 essential video coding standard. In Proceedings of the Applications of Digital Image Processing XLIV, San Diego, CA, USA, 1–5 August 2021; pp. 127–134.

12. International Telecommunication Union. Information technology-digital compression and coding of continuous-tone still images-requirements and guidelines. *CCITT Recomm.* **1993**, *81*, 09.
13. Bjontegaard, G. Response to Call for Proposals for H.26L. ITU-T SG16 Doc. Q15-F-11. In Proceedings of the International Telecommunication Union, Sixth Meeting, Seoul, Republic of Korea, November 1998.
14. *ISO/IEC JTC 1/SC 29/WG 04 N0047*; Report on Essential Video Coding Compression Performance Verification Testing for SDR Content. International Organization for Standardization: Geneva, Switzerland, 2021.
15. Choi, K.; Chen, J.; Rusanovskyy, D.; Choi, K.P.; Jang, E.S. An overview of the MPEG-5 essential video coding standard [standards in a nutshell]. *IEEE Signal Process. Mag.* **2020**, *37*, 160–167. [CrossRef]
16. Park, W.-S.; Kim, M. CNN-Based in-Loop Filtering for Coding Efficiency Improvement. In Proceedings of the 2016 IEEE 12th Image, Video, and Multidimensional Signal Processing Workshop (IVMSP), Bordeaux, France, 11–12 July 2016; pp. 1–5.
17. Bjøntegaard, G. Improvement of BD-PSNR Model, ITU-T SG16/Q6 VCEG-AI11. In Proceedings of the 35th VCEG Meeting, Berlin, Germany, 16–18 July 2008.
18. Dai, Y.; Liu, D.; Wu, F. A Convolutional Neural Network Approach for Post-Processing in HEVC Intra Coding. In Proceedings of the International Conference on Multimedia Model, Reykjavik, Iceland, 4–6 January 2017; pp. 28–39.
19. Kang, J.; Kim, S.; Lee, K.M. Multi-Modal/Multi-Scale Convolutional Neural Network Based in-Loop Filter Design for next Generation Video Codec. In Proceedings of the 2017 IEEE International Conference on Image Processing (ICIP), Beijing, China, 17–20 September 2017; pp. 26–30.
20. Wang, M.Z.; Wan, S.; Gong, H.; Ma, M.Y. Attention-based dual-scale CNN in-loop filter for versatile video coding. *IEEE Access* **2019**, *7*, 145214–145226. [CrossRef]
21. Zhang, Y.; Shen, T.; Ji, X.; Zhang, Y.; Xiong, R.; Dai, Q. Residual Highway Convolutional Neural Networks for In-Loop Filtering in HEVC. *IEEE Trans. Image Process.* **2018**, *27*, 3827–3841. [CrossRef] [PubMed]
22. Wang, Y.; Zhang, J.; Li, Z.; Zeng, X.; Zhang, Z.; Zhang, D.; Long, Y.; Wang, N. Neural Network-Based In-Loop Filter for CLIC 2022. In Proceedings of the 2022 IEEE/CVF Conference on Computer Vision and Pattern Recognition Workshops (CVPRW), New Orleans, LA, USA, 19–20 June 2022; pp. 1773–1776.
23. Huang, Z.; Sun, J.; Guo, X.; Shang, M. One-for-All: An Efficient Variable Convolution Neural Network for In-Loop Filter of VVC. *IEEE Trans. Circuits Syst. Video Technol.* **2022**, *32*, 2342–2355. [CrossRef]
24. Dong, C.; Deng, Y.; Loy, C.C.; Tang, X. Compression Artifacts Reduction by a Deep Convolutional Network. In Proceedings of the IEEE International Conference on Computer Vision, Santiago, Chile, 7–13 December 2015; pp. 576–584.
25. Li, C.; Song, L.; Xie, R.; Zhang, W. CNN Based Post-Processing to Improve HEVC. In Proceedings of the 2017 IEEE International Conference on Image Processing (ICIP), Beijing, China, 17–20 September 2017; pp. 4577–4580.
26. Zhang, F.; Feng, C.; Bull, D.R. Enhancing VVC Through Cnn-Based Post-Processing. In Proceedings of the 2020 IEEE International Conference on Multimedia and Expo (ICME), London, UK, 6–10 July 2020; pp. 1–6.
27. Zhang, F.; Ma, D.; Feng, C.; Bull, D.R. Video Compression with CNN-Based Post Processing. *IEEE MultiMedia* **2021**, *28*, 74–83. [CrossRef]
28. Bonnineau, C.; Hamidouche, W.; Travers, J.F.; Sidaty, N.; Deforges, O. Multitask learning for VVC quality enhancement and super-resolution. In Proceedings of the 2021 Picture Coding Symposium (PCS), Bristol, UK, 29 June–2 July 2021; pp. 1–5.
29. Wang, M.; Wan, S.; Gong, H.; Yu, Y.; Liu, Y. An Integrated CNN-Based Post Processing Filter for Intra Frame in Versatile Video Coding. In Proceedings of the 2019 Asia-Pacific Signal and Information Processing Association Annual Summit and Conference (APSIPA ASC), Lanzhou, China, 18–21 November 2019; pp. 1573–1577.
30. Meng, X.; Deng, X.; Zhu, S.; Zeng, B. Enhancing quality for VVC compressed videos by jointly exploiting spatial details and temporal structure. In Proceedings of the 2019 IEEE International Conference on Image Processing (ICIP), Taipei, Taiwan, 22–25 September 2019; pp. 1193–1197.
31. Shan, L.; Elena, A.; Jonathan, P.; Mathias, W.; Ping, W.; Yan, Y. JVET AHG report: Neural-network-based video coding, JVET-T0011. In Proceedings of the 20th JVET Meeting, Teleconference, Online, 7–16 October 2020.
32. Das, T.; Choi, K.; Choi, J. High Quality Video Frames from VVC: A Deep Neural Network Approach. *IEEE Access* **2023**, *11*, 54254–54264. [CrossRef]
33. Ma, D.; Zhang, F.; Bull, D.R. BVI-DVC: A training database for deep video compression. *IEEE Trans. Multimed.* **2021**, *24*, 3847–3858. [CrossRef]
34. FFmpeg Documentation. Available online: https://ffmpeg.org/ffmpeg.html (accessed on 12 December 2023).
35. EVC Encoder (XEVE). Available online: https://github.com/mpeg5/xeve (accessed on 12 December 2023).
36. EVC Decoder (XEVD). Available online: https://github.com/mpeg5/xevd (accessed on 12 December 2023).
37. Elena, A.; Liao, R.-L.; Liu, S.; Andrew, S. JVET common test conditions and evaluation procedures for neural network-based video coding technology, JVET-AC2016. In Proceedings of the 20th JVET Meeting, Teleconference, Online, 7–16 January 2023.

Disclaimer/Publisher's Note: The statements, opinions and data contained in all publications are solely those of the individual author(s) and contributor(s) and not of MDPI and/or the editor(s). MDPI and/or the editor(s) disclaim responsibility for any injury to people or property resulting from any ideas, methods, instructions or products referred to in the content.

MDPI AG
Grosspeteranlage 5
4052 Basel
Switzerland
Tel.: +41 61 683 77 34

Sensors Editorial Office
E-mail: sensors@mdpi.com
www.mdpi.com/journal/sensors

Disclaimer/Publisher's Note: The title and front matter of this reprint are at the discretion of the Guest Editors. The publisher is not responsible for their content or any associated concerns. The statements, opinions and data contained in all individual articles are solely those of the individual Editors and contributors and not of MDPI. MDPI disclaims responsibility for any injury to people or property resulting from any ideas, methods, instructions or products referred to in the content.